Machine Readable Me

Published by 404 Ink Limited
www.404Ink.com
@404Ink

Please note: Some references include URLs which may change or be unavailable after publication of this book. All references within endnotes were accessible and accurate as of Sept 2023 but may experience link rot from there on in.

Editing: Laura Jones-Rivera
Typesetting: Laura Jones-Rivera
Proofreading: Laura Jones-Rivera
Cover design: Luke Bird
Co-founders and publishers of 404 Ink:
Heather McDaid & Laura Jones-Rivera

Print ISBN: 978-1-912489-82-4
Ebook ISBN: 978-1-912489-83-1

Printed and bound in Great Britain by Clays Ltd, Elcograf S.p.A.

MIX
Paper from responsible sources
FSC
www.fsc.org
FSC® C018072

Machine Readable Me

The Hidden Ways Tech Shapes our Identities

Zara Rahman

Inklings

Contents

Introduction

As we move through the world, information about who we are is gathered from all angles – our fingerprints when we cross borders, our health information when we register at the local pharmacy or doctors' surgery, whenever we post on social media. This information is used to shape the paths available to us in life though we often don't even know that data is being gathered, much less what it's being used for.

In London, UK, children are being entered into databases that affects what happens to them if they get stopped by the police in the future, based purely on where they live, who their families associate with, and the fact they are Black. In India, farmers are being denied access to their pensions, money that they depend on as a matter of life or death, because a machine doesn't recognise their fingerprints, worn down from decades of manual labour. Iris scans taken in Niger travel to the EU faster and with fewer restrictions than bodies of the people they belong to.

Generally speaking, participation in modern life depends on being part of these systems but data that is gathered about us is increasingly limiting what we can and can't do, in ways that we rarely see. These systems have been built to prioritise large-scale efficiency for corporations over personal user experience, giving more weight to what machines say than to what the people themselves say in response.

Machine Readable Me looks at how and why that happens. Based on over a decade of research into how data about who we are is gathered, stored and used by governments and international agencies, and the often-unintentional consequences thereof, my research has taken me to over twenty different countries, where I've had the honour of working with activists, journalists and communities who have been negatively affected by data or technology. I've spoken with refugees who felt they had to choose between giving their fingerprints or gaining shelter for their family for the night, and with people who designed these systems, many with the best of intentions, though that is rarely apparent in operation.

My research tends to be most interested in the margins, in the people who exist outside of the mainstream, or who aren't what a computer or digital system 'expects', communities whose experiences are often ignored, or those who are in a minority. I believe that studying who a digital system *doesn't* work for tells us much more about it than who it does, and I interrogate how power moves within a system, drawing from feminist critiques and feminist methods, motivated by wanting to change social systems writ large to be more equitable and fair.

This topic is personally motivated too, as my own identity doesn't quite fit in the usual given boxes. I grew up in the UK as a British citizen to Bangladeshi parents and as an adult I moved to Germany, becoming a German citizen after the disaster that was Brexit. I've answered the question 'where are you from' more times than I'd like to count, and I've had my own answer contested by strangers an astonishing number of times. I now have two mixed-race children whose own identities are even more between-boxes than my own. They're young but already part of digital systems in both the UK and Germany, and it's astonished me how many decisions my partner and I have had to make about how they should be represented or have data collected about them in the digital world. For me, data and digital systems represent what societies value, and it's clear that something needs to change as social inequality increases and the climate crisis deepens. We all have our part to play in changing that.

Every day, we have to engage with digital systems that seek to identify and/or verify us for different reasons, known as 'digital identification systems'. Most of the systems I talk about here are run by governments or international organisations like agencies within the United Nations. They are termed differently in different countries – the Aadhaar card in India, the Personalausweis in Germany, and more generic identification cards elsewhere. I'll refer to 'identification', the act of distinguishing and/or recognising someone, which usually happens through someone else.[1]

Digital identification systems tend to work by gathering, generating and sorting information which can include a huge range of items like age, nationality, location, financial status, and so much more. For a machine to process information, it is 'translated' into data. Computers understand such data in binary, it's either a 1 or a 0 – there's no grey area, no middle ground. For a computer to process data it needs to be in discrete categories or hold discrete values and, usually, humans decide what those categories are.

For example, when we fill in a form, there's usually a spreadsheet or database somewhere that's populated with the answers – like when you buy a train ticket, giving your information so that your seat is registered in your name. There's no going 'between' the cells or creating new categories to meet everyone's needs. By definition, this kind of data is most useful when it categorises and classifies people, making large populations much easier for states, or those in power, to 'understand', and thus control.

As my work has repeatedly shown me, data can never truly capture a full picture of its subject, because who we are changes depending on context, who we're talking to, where we are, and what we need. We're not the same people that we were ten years ago, nor will we be the same in ten years' time. But as long as systems assume that such data can provide a holistic assessment and play such a crucial role in modern life, they continue to cause harm to those who don't fit the system's expectations. Consider trans people whose gender identities might not fit binary expectations, or climate refugees who decide to leave their homes in search of a better life, whose movement is restricted by the very same people who caused their homes to be uninhabitable. When governments use this data to determine our lives in ways we cannot see, our ability to decide for ourselves who we want to be and what we want to do, is denied. Our self-determination, our autonomy, all overlooked and ultimately called into question, often through egregious violations of human rights.

Technology and data don't exist in a vacuum – they're designed and implemented in societies that are shaped by politics and culture. Issues that might seem far away from those of technology or data but which, as we'll see, are actually deeply intertwined. Structural inequalities, where there are systemic disparities in how power or resources are allocated based not on an individual's actions, but on how the very system is designed and built.

These inequalities can happen when the group of people deciding how resources are allocated either intentionally or unintentionally set the rules so that they discriminate against a certain group. For example, how women are paid less than men, and how within that group, Black women are paid less than white women.

I'll also talk about race, and racial identity within data systems. Race is a social construct, a way of classifying humans that is purely invented *by* humans, rather than anything biological or inherent.[2] Were it not for how society perceives race, it wouldn't be an issue. As Tukufu Zuberi describes in *White Logic, White Methods,* it is 'the international belief in race as real that makes race real in its social consequences.'[3] If we didn't live in a society where being Black or white or brown matters, the melanin in our skin would not impact our lives. But we do. Structural racism is real, and as a result, our racial identity within and outside of data matters.

We are so much more than data gathered. Who we are changes fluidly, and other people's impressions of us should not be codified or set in stone in ways that impact the paths available to us. In many cases, the governmental digital systems that gather data about who we are are built upon problematic policies and laws, the types that discriminate against skin colour, sexuality, sexual preferences, or other aspects of who we are. They cannot be improved without purposeful attention and undoing, without intentional work to unpack whom these policies harm and why. But all too often, data-focused approaches are being used to reinforce those unfair policies and approaches, thereby making structural inequalities stronger.

It's crucial to remember that data is flawed. A digital version of us will never capture our full, messy, fluid selves and when states, companies or organisations try to use those digital versions as proxies for profiling, they're causing harm through misrepresentation and misjudgement that has long, tangible implications.

Chapter 1
How does data try to capture who we are?

In 2012, I read a tweet from the United Nations High Commission for Refugees (UNHCR), one of the world's biggest refugee agencies. The tweet described the reaction of a Senegalese refugee who had just received an identification card, issued by UNHCR.

'At least I have an identity now. I exist.'[4]

The accompanying article went on to extol the ways in which gaining an identification card changed lives, allowing recipients to integrate into local society, gaining access to loans and local schools for their children.[5] It also noted that the cards weren't just simple identification cards, they also held the owners' fingerprints, photo and biographical data. Why, I wondered, would a card need that much information? And what might happen if it fell into the wrong hands? Those musings, and that statement of 'existing' in large part due to that card, stayed with me and ended up being the spark that shaped my research over the coming years.

The way that data is organised, via categories and labels, can have a huge impact on our lives. Journalist Lena Groeger writes, 'decisions about how to design a form have all kinds of hidden consequences', citing many examples of how form design affected important data collection, such as data about different races gathered via the census in the United States, even influencing who people are more likely to vote for.[6] Category creation and curation has always been the source of a great deal of power, long before digital technology spread it faster and further than we'd ever imagined.

In the 1920s, Belgian colonial powers took it upon themselves to institutionalise racial categories that had not, up until that point, played a significant role in Rwandan society. Key to that was carrying out a census that 'classified the entire population as Tutsi, Hutu, or Twa, and issued each person with a card proclaiming his or her official identity',[7] before they moved on to reforming local administration according to these new racialising categories, as Ugandan scholar Mahmood Mamdani explores in his book, *When Victims Become Killers*.

The Belgians, and other colonising powers before them, understood the power of drawing lines in society where there were previously none, and of adding levels of bureaucracy to make life easier for the few that were in power. Under their rule, 'the colonial power constructed the Tutsi as nonindigenous and the Hutu as indigenous ... This had a crucial social effect: neither kwihutura (the social rise of an individual Hutu to the status of a Tutsi) nor gucupira (the social fall from a Tutsi to a Hutu status) was any longer possible. For the first time in the history of the state of Rwanda, the identities "Tutsi" and "Hutu" held permanently. They were frozen.'[8]

Ghanaian-American philosopher and academic Kwame Anthony Appiah calls this the 'Medusa Syndrome', writing that 'what the state gazes upon, it tends to turn to stone.'[9] He describes this inadequate but somewhat inevitable strategy that the nation-state adopts as the only way a state has of making its people legible or, in other words, of 'watching' its population. But watching is not the same as seeing.

Nowadays, our identities are codified in a wide range of systems which are primarily controlled by large institutions through documents like passports and ID cards. States are one of the main players in the identification game, using data to discern those worthy of welfare payments, to charge taxes, to offer health services and pay pensions, and this is nothing new. What *is* new is the digitisation of these systems. With the support and financial encouragement of the World Bank, more and more governments are setting up digital ID schemes that are increasingly tied to welfare and other government-run programs, further integrated with private sector services.

Before we get too far into the problems of such systems, it's worth understanding *why* these systems are established on such a huge scale. Back in 2015, member countries of the United Nations agreed upon a set of goals to shape the following fifteen years, which have come to be known as the Sustainable Development Goals (SDGs). These goals have set a framework for a lot of governmental spending since then and include important (though lofty) targets like achieving zero poverty, zero hunger, and gender equality.

Goal 16.9 states that that everyone should have access to a legal identity – that is, a legally recognised document which provides an official record of the existence of that person and enable the recognition of that person before the law.[10] Right now, millions don't have that, for a variety of reasons. They might not have access to a mechanism to register births, which means that when they are adults they can't officially prove that they were born. Or they might not be recognised as equal to others in their country simply because of the family they were born into, or the conditions of their birth, and as a result, not be entitled to a legal identity document. This often happens to people born into ethnic minority communities who are discriminated against by those in power – as we'll see shortly, communities like Nubians in Kenya, or Rohingya in Myanmar.

The issue of access to legal documents is important enough that in June 2023, stateless and undocumented people living in the Kavango East and Kavango West regions of Namibia held a demonstration to call on the Namibian government to speed up the registration process.[11] The group flagged the irony that death certificates of undocumented people are processed faster than documentation to help stateless and undocumented people live their lives.

Theoretically, digitising this process should mean registration happens efficiently, and be easily recorded and accessed at scale, potentially addressing many of the hurdles towards providing legal identity documents to people. But this is a fallacy. Legal identity is often withheld by governments not for a lack of proper system, but because of societal bias or discrimination. Any digital identification systems that are established without actively addressing those biases will simply replicate that discrimination, thus not really addressing the core issue at all.

We can see this kind of discrimination replication in how members of the minority Nubian community in Kenya have been treated. Their ancestors were forcibly brought from Sudan by the British colonial government over a century ago, and despite there now being over 100,000 Nubians in Kenya,[12] they've faced discrimination ever since, with the Kenyan government denying them their rights as citizens of Kenya. They've struggled to access identity documents which are 'vital tools in fighting for other rights such as education, land rights, seeking loans and employment opportunities.'[13] Organisations like the Nubian Rights Forum have been doing crucial work helping Nubian people access identity documents in the face of that discrimination, via radio talk shows on citizenship and related issues, and legal support, aiming to empower Nubian community members to know and claim their rights.

In January 2019, the government of Kenya launched a national, mandatory digital identity card system, popularly known as Huduma Namba (meaning 'service number' in Swahili) which would be a prerequisite to access vital social and governmental services. To enroll for Huduma Namba, though, identification documents are needed, documents that many members of the Nubian community struggle to access in the first place. Human rights groups such as the Nubian Rights Forum, the Kenya Human Rights Commission, and the Kenya National Commission on Human Rights have launched multiple legal challenges against the proposed system.[14]

That's not the only issue with the Huduma Namba system as it was proposed, but it does highlight just one way that digital systems can unintentionally facilitate discrimination. There was enough resistance to the system that in early 2023, the budget for the Huduma Namba was cut by 84%, effectively ending the rollout.[15] Uptake was low and rights groups carried out successful legal challenges against the system. Even in the face of well-funded, nationwide systems like the Huduma Namba, well-organised resistance and protest can make a difference.

Digital identification systems such as the Huduma Namba aren't necessarily a part of reaching Goal 16.9, despite its appearance. Legal identity documents are important because they enable the recognition of that person before the law, which facilitates the claiming of other rights. But many digital systems are a prerequisite to accessing government services, as with Huduma Namba – meaning that their actual use is far beyond simply providing legal identity. Often these other government services involve some level of involvement with the private sector, too. When that's the case, they often gather much more data that wouldn't at all be necessary for simple legal identity. For example, the Huduma Namba was originally proposed to include DNA, retina scans, iris patterns, voice waves and even earlobe geometry, all to be stored with personal information in a central database.[16]

Digital identification systems are often proposed as a 'one stop shop', a single identification system to facilitate access into other systems, whether government-run or private sector. This is not what the SDGs were aiming for but it does meet certain goals of states, as political scientist James C. Scott described in his 1998 book, *Seeing like a State: How Certain Schemes to Improve the Human Condition Have Failed.*[17] He notes that states (and other large institutions) gather data on large populations to make them 'more legible.' In his eyes, the driving purpose behind gathering all this data isn't necessarily in the best interests of the people whose data it is, but rather for people in power to observe who they are governing or ruling over, to help *themselves* to make decisions as according to their values and priorities, not their citizens.

Gathering data about who *we* are, should be in *our* best interests. Sure, states need certain information to run efficiently and manage populations at scale, but the type of data that we're talking about gathering in these identification systems is deeply personal, and, as we'll see, will sometimes follow us through our lives.

Chapter 2
The assumptions within our data

Social, political and cultural assumptions go into the creation and shaping of information. Information like our country of origin, and our nationalities. For example, how many countries actually exist? Who decides what makes a country a country? For many, it's United Nations recognition that makes a country 'official.' In some cases, countries only exist in that shape or size because colonisers decided where they should be, who governed them and how, violently massacring or removing Indigenous people from what was rightfully their land. Maps show us where artificial partitioning happened – straight lines across the continent of Africa, borders with perfect right angles between Sudan, Libya and Egypt, to name just a few.

Some of those lines were drawn on a map just a few hundred metres from my first apartment in Berlin, on Wilhemstrasse. The Berlin conference of 1884 is where the white men of Europe came together to divide up Africa, though it was well under way before the conference itself. Representatives from every European country except for Switzerland, as well as the United States, were present. There were no African attendees. Author Patrick Gathara describes that the conference was where 'the rules for the conquest and partition of Africa' were established, 'legitimising the ideas of Africa as a playground for outsiders.'[18]

Regardless of when exactly the lines were drawn, many of them remain today. It's wild that global society has largely agreed to maintain those lines, regardless of inhabitants needs or desires. But there are exceptions. Border disputes do occur from time to time, and it's here where the collection of digital data makes us confront the lines and assumptions that come with them.

As of 2021, 139 countries had bilaterally recognised and acknowledged Palestine as a country, but at the time of writing in August 2023, certain rich countries, including the United States, Japan, France and Germany, still do not. Unless you're from countries whose land is disputed, you probably wouldn't notice it, but when data is being gathered about your country of origin and your country is deemed invalid, it can hurt on an existential level, as well as a physical one. For Palestinians, it means that in certain circumstances like applying for a visa to visit the United States, the closest option they're offered for their country of origin is Israel, a country that has violently occupied their land for decades. Not only has the Israeli government physically occupied Palestinian land, but that occupation exists digitally too, co-opting their identities and leaving them with no option but to select the nationality of their occupiers instead of their own.

Broadly speaking, societies have chosen to make nationalities and countries of origin vital in how we move about the world, and in the opportunities available to us. This way of organising society is the result of intentional decisions made by the few. We consider 'foreigners' to be people from outside of the country that we are referring to but we also know these borders, being 'foreign', are manufactured. In a book entitled *The Invention of the Passport,* author John Torpey describes how, as late the nineteenth century, people coming from neighbouring provinces in France were considered just as foreign as people from other countries.[19]

It wasn't until after World War I that the concept of a worldwide passport necessary for travel was suggested, which put more emphasis on a person's country of origin. This system was designed with the large numbers of displaced people in post-war Europe in mind: passports gave a way of monitoring and tracking who was allowed to enter and leave different territories. While mainly a tool for movement, these passports also gave people living in the same country a shared identity in a way that was new, centred around their citizenship for their nation-state.

Nowadays, thanks to the strengthening of sovereign nation-states, our nationalities have an outsized importance in how we relate to and move in the world. These labels that we grow up with, determined by the luck of the draw, combined with governmental and institutional powers affect how we behave, how we see each other and ourselves. Entirely unfairly, owning a passport for a rich or developed country grants us far more freedom of movement and a stronger claim to certain other human rights too, than not owning a passport, or owning a passport of a poorer country. With a German passport, I can visit 124 countries without needing a visa – but with a Bangladeshi passport, I can only visit 14 countries visa-free. It's purely luck that I have one and not the other. Most of my cousins and family are reliant on time-consuming, expensive and frustrating visa application processes while I can book tickets and hop on a train without a care in the world. It's deeply, deeply unjust.

Turkish author Orhan Pamuk wrote in the *New Yorker* in 2007, 'a passport is not a document that tells us who we are but a document that shows what other people think of us.'[20] It's easy to forget how incredibly artificial and man-made those borders that influence passports – and these documents, or digital ID systems – really are. In the digital world, the answer to where we're from is synonymous with what passport you hold. But being *from* a certain place is about much more than one's identification document. Being from somewhere can mean feeling grounded to that place, feeling a sense of belonging and familiarity, recognising the smells of the food or the sounds of the language. Syrian American writer Zeyn Joukhadar writes, 'When this country asks me where I'm from, they aren't asking for the city on my birth certificate, but whose earth is in my blood.'[21] Belonging has cultural and social aspects that cannot be quantified or measured, and the flatness of digital data capture doesn't allow for that level of complexity.

This one-dimensionality rears its head in ancestry testing, or genome testing, where companies like 23andMe offer insight into a person's place(s) of origin based on genomes. You spit in a tube, post it, and they send you your results in a few weeks. Sounds simple, but behind the glossy branding is much messier and complicated service than they'd like you to believe.

Commercial genome tests look at DNA and compare it to samples held in its database. Within this database, reference groups (groups of people who are known to be from a specific area) are used to estimate markers as being more positively correlated with populations from a certain geography. Reference groups tend to be self-reported, which means there are some limitations in their accuracy. Different companies use different reference groups, and as these reference groups grow (as the data the companies hold grows and changes) the results that are produced may also change retroactively.

Because users of these primarily US-based companies offering genome testing are not globally representative, some reference groups are bigger than others, meaning the bigger ones have higher accuracy. For example, if the database is highly populated with samples from the same area, it is easier to match a new customer to that area. If there are very few samples from an area such as the continent of Africa, the most that a company might be able to say is that someone is from that broad region with little further specificity. Typically, people from the continents of Asia and Africa are under-represented, so the results offered for people of Asian and African ancestry is of lower accuracy. Ancestry can also be more difficult to interpret for formerly colonised or enslaved populations because European colonisers moved forcibly moved thousands of people. This kind of behaviour is typically not considered with the creation of reference groups.

For people who don't get the results they're hoping for, indicating they're not in fact 'from' the place they thought they were, the outcome can influence relationships and the sense of self. A 2020 research study showed that for people with low levels of knowledge about genetics, conducting genetic ancestry tests increased their belief in essentialist views of race as being fixed and 'determining innate abilities,'[22] that is, laying the groundwork for racist and stereotyping beliefs.

It might therefore be unsurprising that white supremacists are big advocates of ancestry tests, keen to have their identity categorised and bolster their racism. As researchers Aaron Panofsky and Joan Donovan described in their 2019 paper, *'Genetic ancestry testing among white nationalists'*, it's not uncommon for those seeking membership of white supremacist movements to be greatly perturbed by ancestry test findings that they are not as 'white' as they had thought.[23] Because of white nationalists' belief in genetic essentialism, sharing this information with fellow white nationalist peers can result in exclusion from the community, and identity crises (or, in some cases, a push towards anti-science beliefs).

Much has been written about ancestry testing challenging our understandings of identity and community groups, not just among white nationalists. Journalist Angela Saini writes in her book, *Superior: the Rise of Race Science* that 'by forcing people into categories, even if that means dividing our individual bodies into so much European, so much African, so much Asian and so on, the tests fortify the assumption that race is biologically meaningful. If it's possible to categorise, we assume, there must be something to the categories.'[24] But as we know, that assumption is incorrect.

Various genome testing message boards indicate very real impacts on the way users see themselves, regardless of the (significant) disclaimers as to the results' accuracy. Knock-on effects can include people choosing baby names based on the ancestry that they discovered via a test, when they have little to no connection to said ancestry or culture; people changing the way they talk, or the way they dress, or the authority with which they speak about topics that they actually don't know much about. It's their test results which make them feel emboldened to have strong opinions on a huge range of issues, ignoring the complexity of belonging or affinity to a certain place or culture. It's yet another example of how digital data about us can misrepresent a slice of a wider story, never the whole picture – and of how physical borders have made their way into the digital space, whether we like it or not.

In 1996, John Perry Barlow declared ‹our identities may be distributed across many of your jurisdictions' in his seminal 'A Declaration of the Independence of Cyberspace.' In actual fact, all we've ended up with is more fixation upon our borders, translated into online systems. We're not distributing identities or ignoring them, let alone creating new ones where everyone is equal no matter where they're from.

Chapter 3

Who do they think you are? Categories, classification and profiling

As an immigrant twice over, first through my parents who moved to the UK from Bangladesh half a century ago, then through my own decision to move to Germany in the early 2010s, the feeling of not quite fitting into people's expected boxes, or the given categories on bureaucratic forms, is quite normal for me.

Sometimes it's left me unsure what the most accurate answer is for seemingly simple questions. Drop-down menus with the question 'Where are you from?' leave me baffled. An imprecise question, usually with a specific and hidden intention behind it. If it's in a digital system, there's no way of asking what that specific intention actually is.

My answer? The UK, I suppose, though I haven't lived there for over a decade. It could be Germany, as it's the place I've spent most of my adult life. If this interaction was to happen in person, at least in Germany, I rarely get to say either of those options without some gesture towards my skin colour, a single eyebrow raised with 'no, where are you *really* from?', exposing a stunning ignorance of colonialism and migration. Borders change, but society's insistence on classifications and categories rarely keep up.

But these problems are somewhat trivial when we consider what category deviation can do to people's lives. In India, 2010, the government introduced the Aadhaar card, a mandatory digital identification card which provides every Indian citizen with a unique ID number which can be used for proof of identity, address and more. The Aadhaar database is currently the largest database of biometric data in the entire world, with over a billion people represented in it. This type of digital ID system, one unique number to be used for multiple government and private sector services, served as an inspiration to Kenya's proposed Huduma Namba card, which I discussed earlier.

The general idea doesn't sound too bad, as many of us struggle with remembering logins and passwords, or being unable to transfer data from one government service to another. When my mother moved house in her 70s, it astonished me how many different places she needed to update her address. If she had had one unique number for everything related to government services (and, increasingly, also private sector services too), it's easy to imagine that those systems would be much more streamlined and efficient for everyone involved.

But of course, it's not that simple. Friction is (and, perhaps, should be more often) purposefully built into digital systems – making it harder to easily transfer data from one system to another as a security feature, or adding in a moment of pause to allow the user to double check that they're doing what they intended, for example. Researcher Anja Kovacs writes about the myriad of harms that have arisen from Aadhaar, including its connection with digital wallets leading to the identities of sex workers being revealed, which is a problem not just for general stigma but also because both 'soliciting and living off the income of sex work are crimes' in India.[25]

Using just one system for everything can have problems, too. In 2017, Indian media outlet *The Wire* reported that there were at least 25 hunger-related deaths directly linked to issues with the Aadhaar card.[26] Even though 64-year-old Premani Kunwar had 'authenticated' herself using the Aadhaar system, her social security pension was paid into someone else's bank account and she couldn't get it changed. 80-year-old Khetrabasi

Pradhan died because his age was incorrectly recorded on his voter card, so the system classified him as not being eligible for a pension when in fact he was.[27]

The way that we are categorised matters –deciding what gets measured or counted and what gets ignored are all ways of exerting power over populations. Philosopher Kwame Anthony Appiah says, 'once labels are applied to people, ideas about people who fit the label come to have social and psychological effects. In particular, these ideas shape the ways people conceive of themselves and their projects.'[28] And in this way, the power that comes from deciding on labels and categories goes beyond simply making populations legible or controlling how populations are treated, but it also affects how we think about each other and ourselves. That's a huge power to hand over to anyone.

For the Rohingya, an ethnic minority population who are indigenous to Myanmar, the right to self-determine their classifications has been a question of life and death. The Myanmar government has tried over decades to erase the Rohingya from Myanmar, starting by changing citizenship laws to no longer recognise the Rohingya as one of the 135 'national races' of Burma in 1982.

This bureaucratic violence[29] was combined with physical attempts at genocide, forced labour, and destruction of property and homes, with the largest genocide taking place as recently as 2017. For decades, the Myanmar government has refused to use the term 'Rohingya', instead describing them as 'Bengali'. Even when defending her country's government against charges of genocide at the International Court of Justice in 2019, then-head of state Aung San Suu Kyi spoke for thirty minutes and did not use the word 'Rohingya' once, preferring instead to describe them as 'Muslims' or 'members of Rakhine communities.' Rohingya activist Wai Wai Nu told Al Jazeera at the time 'refusing to use the term Rohingya means she still doesn't acknowledge the root cause of the genocide allegation.'[30]

After the genocide, hundreds of thousands of Rohingya fled Myanmar and ended up in neighbouring Bangladesh, where one of the world's largest refugee camps stands today. When there's a sudden influx of people on the move or refugees to a new country, head counts are essential, so that humanitarian and governmental agencies know what kind of support the community might need. When the Rohingya arrived in Bangladesh, the UNHCR agreed with the government of Bangladesh to carry out the counting exercise together, then issue identification cards to the Rohingya.[31] But once again, the cards did not acknowledge their Rohingya identity. Instead, they were identified as being from Myanmar. On some identification documents from Myanmar, they were labelled as being from Bangladesh. No official documents acknowledged their ethnic identity as what they wanted it to be: Rohingya.

In November 2018, Rohingya communities in refugee camps at Cox's Bazar in Bangladesh went on strike, with their demands shared on social media:

'We are doing this to demand:

1. Must stop forcing Rohingya refugees to take the Smart Card
2. Must stop barricading Rohingya inside Camp 21 due to their refusing to take the Smart Card
3. Must include our ethnic name 'Rohingya' on the Smart Card, not just 'Forcibly Displaced Myanmar National'
4. Stop the collecting of our biodata and do not share biodata already collected with the Myanmar Government [sic]'[32]

Their strike was even more notable within the broader context of collecting or labelling people according to ethnicity because there's a strong precedent in many countries against collecting or documenting ethnicity data – for good reasons, as we'll see later on in this chapter. In contrast to that precedent, this strike, and the Rohingya's demands, were so clearly connected to their context, given that they had been forced to flee their home country and had their identity and personhood denied for so long, both bureaucratic and explicitly violent.

Intentionally refusing to allow members of marginalised communities to identify themselves as they see fit is a technique used by states all over the world to oppress and silence groups. Speaking to Maya Ch'orti' and Zapotec environmental scientist Dr. Jessica Hernandez, Maya Mam activist Juanita Cabrera López notes that 'our Indigeneity as Maya people does not appear or is reflected in our birth certificates. It does not appear in our IDs... It is really in the interest of states to deny our existence and to deny our identity.'[33] But when context differs, the very same type of data – ethnicity data – can play a very different role. This is where acts of protest and resistance can include actions like destroying data.

In the Netherlands, prior to Nazi occupation, a 'comprehensive population registration system for administrative and statistical purposes'[34] had already been completed, which included the collection of ethnicity data, notably flagging who was Jewish and who was not. This structure, aimed at following people 'from cradle to grave',[35] was then adapted to create special registration systems covering Jewish and Roma populations within the Netherlands, and 'played an important role in the[ir] apprehension.'[36]

Forging identification cards became a vital act of resistance against the Nazis, many created by the

'Persoonsbewijzencentrale' (Identity Card Center, PBC), which created an estimated 65,000 forged cards.[37] But as the identification cards were matched against records in the civil registries, it wasn't enough to carry a forged card as it could easily be proven fake. So, in 1943, the Dutch resistance, including members of the PBC, carried out a complex and daring operation to bomb the Amsterdam civil registry office to prevent Nazis from continuing to use the registry to identify and persecute Jews. Though they weren't entirely successful in their efforts, they managed to destroy 15% of the records (800,000 cards), likely saving the lives of many people. Resistance members were even helped by sympathetic fire fighters who intentionally delayed deploying fire engines to give the fire as much time as possible to spread – and they sprayed water liberally to destroy the records, too.[38] Twelve members of the group who carried out the bombing were executed in 1944.

Despite these incredibly courageous efforts, Dutch Jews had the highest death rate among all other occupied western European countries, at 73% of its population. As comparison, the death rate among the Jewish population in Belgium was 40% and in France, 25%. Dutch digital rights activist Hans de Zwart notes there are many, many lessons about privacy rights that we can learn from this tragic period in history – including that personal data can obviously be used for nefarious means.[39]

These examples demonstrate how deeply context-specific identity-focused data is. The broader context around *who* is doing the data collection, *how* the data is collected, and *who* decides its use can be the difference between life and death, without exaggeration.

What might be a fight for self-determination for one community, and a fight that they're willing to forego food and water for, might be a target on the backs for another community, one that they want to get rid of at all costs. Solidarity is needed in all such situations, recognition of the differing forms of oppression at play, and support in resistance, even if that support manifests itself in drastically different ways in different contexts.

During the COVID-19 pandemic of 2020, analysis by English government body Public Health England entitled 'Disparities in the risk and outcomes of COVID-19' revealed that people of Bangladeshi ethnicity were twice as likely to die of Covid than people of white British ethnicity.[40] And indeed, throughout the three waves of the pandemic (thus far) in the United Kingdom, the rate of death involving COVID-19 has been highest for Bangladeshi people than any other ethnic group, while all ethnic minority groups face higher risks than white British people.

Learning this hit me hard, as my entire family is of Bangladeshi origin. To know that they were more likely to be at risk was gutting to read. Even worse was the realisation that nobody knew precisely why. Was it embedded inequities within the healthcare system itself? I knew of research in the United States that revealed Black people's pain is generally taken less seriously and that there's significant and documented racial bias in pain treatment recommendations.[41] Or was it due to other factors?

Report after report confirmed this pattern. The Office for National Statistics released corroborating data,[42] and *The Lancet* called for ethnicity and COVID-19 to be an 'urgent public health research priority.'[43] But British race equality and civil rights think tank The Runnymede Trust noted that it wasn't *just* that health and ethnicity were deeply intertwined. 'COVID-19 is not just a health crisis; it is also a social and economic crisis…and the ability to cope…is vastly different for people from different ethnic and socioeconomic backgrounds.'[44] In their research they found that Black and ethnic minority people were more likely to be classed as 'key workers' in the UK; to be living in a household with more people; to be working outside their house during the pandemic; and to have experienced financial strife due to the pandemic.[45]

Being able to see these patterns was only possible because of disaggregated data – and that was only possible because of the ethnicity categories that were available for people to choose from in the Census in England. Those categories and the way they're presented differs greatly across different countries and even across different governmental services offered in the United Kingdom.

I was deeply grateful, perhaps naively, that this awful pattern could be identified, though I remained largely unconvinced that this discovery would lead to the systemic change necessary to properly address it. But here in Germany, the site of some of the worst offences of historical data misuse, ethnicity data is now rarely collected. This lack of data means that inequities based on race are hard to prove, assess and understand, and even harder to address as a result. In 2020, a Berlin-based NGO based called Each One Teach One took it upon themselves to address this dearth of data. They set up the Afrozensus[46], a project that asked people of African origin living in Germany for more information about themselves, so that this often-overlooked community had some quantitative data to use when advocating for their needs and rights in policy debates.

It's because of data collected about ethnicity that important insights demonstrating race-based inequities can be proven. Without that kind of data, it would be impossible to say how people of different ethnicities were affected by Covid, for example. So, perhaps counter-intuitively, the desire to protect ethnic minorities by not collecting ethnicity data can actually end up harming them, leaving minority groups in countries like Germany unable to prove that they are being systemically discriminated against. That is starting to change among some communities though, as science journalist Hristio Boytchev notes in *Nature,* 'cultural unease [about collecting data on race or ethnicity] is starting to shift.'[47]

It means that collecting this kind of data is, if done right, an act of empowerment, somewhat similar in goals and motivation to the Rohingya. Until race- or ethnicity-based discrimination has been entirely eliminated, there's a clear need for this kind of data to be collected – to be able to highlight where those inequities are, and, ideally, to suggest interventions and systemic changes that can be made to address them.

If and when this kind of data is collected, though, the structures of accountability and trust around it really matter, as well as *how* it is collected and under what terms. Databases of such sensitive information can easily fall into the wrong hands and be used to persecute instead of empowering. We've seen this throughout history, from the 1939 Census in Germany including a supplementary card inquiring if a person's grandparents is or was a Jew, along with the assurance that 'anonymity would be respected',[48] which in fact went straight to a department run by Adolf Eichmann to 'close gaps on in their existing card index on Jews',[49] breaking that anonymity immediately, to biometric data collected by Western governments falling into the hands of the Taliban in Afghanistan in 2021.[50]

In short, respecting people's self-determination and autonomy when it comes to sensitive data about who we are is complex and hard to do well. But ignoring that kind of data is also not an option.

As activist and professor Ibram X Kendi explains, 'terminating racial categories is potentially the last, not the first, step in the anti-racist struggle.'[51] It really matters who who decides what the categories are, how that data is gathered, who holds it, and if and when it's deleted. The data we need is shaped by the society we live in, which have racial injustices built into their very core. We need data about ethnicity to make those injustices visible. But as ever, just having the data is not enough.

Chapter 4
Growing and changing – or not

Identity fluidity is a core part of humanity. As Octavia Butler writes, 'the only lasting truth is change.' You're probably not the same person you were ten, or even five, years ago. In fact, nowadays, social media offers us opportunities to see how true this is, just by looking at what we wrote in old posts or in the photos we posted multiple years ago. If I look back at my old Facebook account, I cringe at the status updates I wrote in the early 2000s, or the captions I wrote on photos. Unsurprisingly, 18-year-old Zara had significantly different tastes and preferences to Zara today.

With this in mind, it can be frustrating when friends or acquaintances don't acknowledge how much we've changed or grown, or worse, use it against us. That can be at the heart of disputes between friends who haven't seen each other in years, or family members that refuse to see how different you are from the little child who used to run around at hectic birthday parties.

Even at a particular point in time, identity presentation is fluid. Who I am at any given time depends on a confluence of factors and the context of the situation I'm in. Sometimes I might want to perform one element of my identity more than another, depending on who I'm with and what's most important to me at that particular time. That choice is, or should be, mine to decide.

For states and institutions wanting to understand their populations, almost the exact opposite to Butler's quote is true. According to digital data, collected at a particular point in time, and entered into a system, who we are is static, a series of unchanging, cold facts. We are categorised by where we live, our gender identity, our year of birth, perhaps by our ethnicity, and increasingly, systems we interact with online categorise us by our online behaviour, our digital traces. We're rarely aware of how these systems are 'seeing' us, which makes adjusting them or changing them nigh impossible.

The move to digitising our identities, or digitising more data about our identities, makes this fluidity much harder. Digital data adds a rigidity to our identities as far as machines are concerned, a permanence, forcing a framework upon a personhood that we might not agree with, let alone personally identify with. This can be really problematic for many reasons. For one, the assumption that this rigidity is at all compatible with the way that people change over time. That people's social networks would by default stay the same over time, that their approaches, views and preferences would stay the same instead of change, makes it seem like growing and changing is unusual instead of the norm. Changing your mind (or your preferences) is often a sign of having learned new information but if our technical systems insist that your preferences from years ago are the same ones that dictate what information you should get today, then you're probably less likely to, firstly, be exposed to new information, and, secondly, be able to change those preferences to make them match who you are today.

In some ways, society does the same. In some countries, people who have been incarcerated are forced by law to always mention this in job applications, which can drastically reduce their chances of getting a job. This further emphasises the uselessness of the 'overlapping interests of government and industry that use surveillance, policing, and imprisonment as solutions to economic, social and political problems' which abolition group Critical Resistance call the 'prison industrial complex.'[52] Under the current systems in place, there's no chance for redemption and no chance for forgiveness in the eyes of many, let alone in the digital (and analogue) systems built around them.

Systems like credit scoring or criminal records create profiles that stay with individuals, regardless of whether they've changed as people. Once categorised with a certain label, that label sticks, no matter how much a person's identity or behaviour changes. Worst of all, participation in these systems is not optional, yet they purport to increase efficiency or improve said systems. Sometimes, inclusion in certain databases can happen without people even knowing, with disastrous consequences.

The Gangs Matrix is one such system, a police database set up in the United Kingdom by the London Metropolitan police in 2012. It includes 'suspected gang members in London', but as Amnesty International revealed in a damning report in 2018, it has ended up being 'a racially discriminatory system that stigmatises young black men for the music they listen to or their behaviour on social media.'[53] A disproportionate percentage of members of the database are Black, and people who have never actually been involved in violent crime are also registered in the Matrix. Despite the deep subjectivity of being included in this database, it's shared with partner governmental agencies, and thus 'stigma of being associated with a "gang" can follow individuals through their interactions with services such as job centres, housing associations and schools, and may result in negative outcomes.'[54]

The very label of 'being in a gang', and the assigning of such an identity to people, brings a deep stigma with it in today's political climate, while there's actually a very blurry definition of what being in a gang actually means. It's also, as the Amnesty report explains, a non-solution to the actual problem of knife crime in London. A report released by the Mayor of London's office in 2017 noted that 'recent data suggests that the majority of knife crime is not gang-related. Gang-flagged crime accounted for 5% of all knife crime with injury during 2016'[55] – thus suggesting that the very approach of attempting to focus on 'gangs' as a way of addressing knife crime in the UK's capital is deeply flawed.

People aren't notified in any way when they're added to the database, and knowledge of the very existence of the database itself is not wide. This secrecy makes it impossible for people to object to how they are being categorised and how their data is being recorded, and the Matrix database being shared with other agencies without explicit consent from the people involved means that their lives could be being affected in all sorts of ways without them even knowing. Job applications rejected at the first round without any good reason, or child protective services turning up without any warning.

And as ever, it's people who are already marginalised

who are most negatively impacted and discriminated against by such systems. This trend isn't unique to the UK, either. In 2018, a research collective named Our Data Bodies carried out a project examining 'the impact of data collection and data-driven systems on the ability of marginalised people to meet their human needs.'[56] They focused on three different locations within the United States, and concluded that the systems they looked into were 'for the sole benefit of the data collector,' rather than for community members themselves.

Sadly, this structure, where the aims and goals of a system are for the benefit of the people who designed the system but involve the data collection or active involvement of other people, is a relatively common one that we'll see again and again throughout this book. In the case of the Gangs Matrix, the database served as a political response to the 2011 riots in London, demonstrating that there was *a* response, that it was tech-savvy and could easily be spread and scaled. But it was the wrong response, a non-solution to the problem, not to mention a deeply discriminatory system. Once again, a tech-first solution was prioritised above actual engagement with the structural issues causing the problem.

Tech critic Evgeny Morozov refers to this perception as 'technological solutionism',[57] the idea that every social problem can be solved by a technical solution, and that we've built a reliance upon (and myopia towards) the unreliability and human bias built within technical solutions.

Using digitised systems for decisions like this adds layers of perceived legitimacy to the decisions behind them. Let's look at the Gangs Matrix again. The Met Police describes it as an 'intelligence tool' which 'measures the harm "gang nominals" (who are named on the matrix) pose by scoring them based on evidence of them committing violence and weapons offences, any police intelligence relating to them having access to weapons, or them being involved in, or at risk from, gang violence.'[58] The words in that description – intelligence, scoring, evidence – all indicate a level of objectivity, of infallibility in a way, which makes it harder to scrutinise or to see where mistakes might be made.

But it's not just through the creation of databases or digital systems that technology can play a role in cementing an identity that we might not want or deserve. Pieces of data that we generate ourselves can also be misinterpreted to build a picture of us that we might disagree with, thanks once again to a combination of misunderstanding of technology, human bias, and technical illiteracy.

In Manchester, in July 2022, four Black boys were sentenced to eight years in prison for 'conspiracy to cause grievous bodily harm.'[59] The main pieces of evidence that convicted them were messages they sent on messaging app Telegram. Despite that none of the boys themselves were found to have had any weapons, or have actually taken part in any violent acts, the jury found these four guilty of taking part in a conspiracy. One of the boys, Ademola Adedeji, was only active in the chat for fourteen minutes, passing on information that was never acted upon, and based on which no crime was actually committed.

A key part of this case rested on the prosecution (and eventually the judge) deciding that the boys were part of a gang, while the boys themselves insisted that the group that some of them were part of was actually a music group. And even then, some of the defendants weren't even in the group, but had simply watched a couple of their videos on YouTube.

In this case, digital data about their actions on their phones was used to make inferences about the boys' characters and identities, to try to 'prove' that they were guilty not even of an actual crime, but guilty of the *intent* to commit a crime. What we're seeing here is how digital data is being used to violate their right to freedom of association, of freedom of expression, and, fundamentally, of privacy too. It was used to put together a picture about who those boys were, regardless of who they *actually* are – whether that's understood as who they say they are for themselves, or even who those close to them think they are. The fact that they were in a group chat with specific other people, is interpreted as them 'conspiring' together. The fact that they were in a music group with others, who produced videos and music together, is interpreted as them being in a gang together – an interpretation which would not be true if the boys were not Black.

Besides this being yet another example of deep-seated anti-Black racism and injustice within the British criminal 'justice' system, it shows us how important having self-determination is, the ability to decide what information is most important about us. It didn't matter that Ademola Adedeji was Head Boy at his school, had a place to study Law at Birmingham University, nor that he was a model student and key member of his community. The pieces of data that were paid attention to were those which (supposedly) associated him with 'gang' culture, and messages he sent over twenty minutes.

As documented extensively by Manchester-based advocacy project Kids of Colour[60], he was given no opportunity to demonstrate or show that he'd written those messages in a flash of anger, and that those particular data points were not in fact a good representation of who he was as a full person. What was 'noise' – an anomaly, a one-off – was used as a signal pointing towards his true character. The consequences are devastating. Instead of taking up that place at university, he will be spending the next eight years in prison, and even his right to appeal has been denied. Thanks to those pieces of digital data combined with the misinterpretation and systemic racism at play, his entire life has been changed forever.

But even though I mention Adedeji's outstanding school record as a way of pointing out how utterly farcical the whole case is, the boys shouldn't have to be outstanding students or especially beloved members of a community for their basic rights to be respected. If digital data is to be used in court cases like this, it must be accompanied by appropriate inclusion and acknowledgement of people's lived experiences as well – including the context in which the data was created. By ignoring that context, digital data can be used to put together pictures of people that will purely further existing biases and discrimination, as we can see in this case.

For some people, growing and changing your identity can have big ramifications on how you live your life. Trans people, for example, might need to update their digital profiles or their records on digital identification systems to more accurately represent who they are post-transition, and allow them to live the lives they choose to live. As Human Rights Watch notes, 'legal gender recognition is fundamental to transgender people's human rights and well-being,'[61] with consequences of identity documents not matching gender classifications including discrimination in accessing healthcare, finding employment, and increased scrutiny for everyday activities.[62]

In many countries though, the rigidity built into identification systems and associated policies deliberately makes this impossible. In Mexico, for example, trans people are required to litigate their own identities due to a lack of proper gender recognition policy, something that activists are trying to change. Research indicates that having some form of ID that matches a person's gender might lead to better mental health outcomes, with early research showing that rates of suicide ideation are lower for people who do hold an ID that matches their gender, in comparison to transgender people who don't hold such an ID.[63]

Progressive policies around digital identification are, slowly, beginning to pop up. Since September 2015, in Ireland, trans people can apply to have their preferred gender legally recognised by the State, via a Gender Recognition Certificate which is then used to provide a revised birth certificate.[64] In Uruguay, transgender people can self-identify their own gender on official documents, and change their legal name without needing approval from a judge.[65] It's unclear, though, whether that self-determination to change gender also means that their past records would be erased, or similarly adjusted to reflect their choices. These are, sadly, the exception not the norm, but it does demonstrate what a system that does put self-determination front and centre can look like.

Hopefully, all these cases show how deeply political data collection is, and how, if taken out of context, technology can have devastating effects on people's lives. We need to have control over data about us, to be able to exercise our right to self-determination on an ongoing basis. This means being able to change data about us when parts of our identities change, or being able to remove data about us if it's painting a false picture of us that we don't agree with. This kind of fluidity doesn't come naturally to digital systems, at least not most of them – which is why it needs to be explicitly built in or designed for if it is to be present at all.

Chapter 5
Data from the body: biometrics

When I became a German citizen a few years ago, I became entitled to a national identity card – a *Personalausweis*. All German citizens hold one and it can be used for a variety of different things, including travel within the European Union, but also for more mundane things like confirming your identity at the Post Office when picking packages up, or proving your name when you want to open a bank account. When I went to the *Bürgeramt (*administrative office) to pick up my card for the very first time, I was asked a few questions by the civil servant working there, one of which took me by surprise. She asked, 'would you like to include your biometric data on the card? Your fingerprints?' Knowing how biometric data can be misused, I took a beat before answering, and asked her in return, 'why? What are the benefits?'

She looked at me for a second, then threw her hands up.

'Honestly, nothing really. I wouldn't if I were you, and I haven't myself. It's more a case of, maybe one day in the future there will be a use for biometrics on the card, but for now, it doesn't do anything. Safer to keep it off there if you ask me!'

She was right. Keeping my fingerprints off the card was the safer thing to do. If there's no reason to collect the data, then the data shouldn't be collected, just in case. But two things struck me: that her priorities were not to try to collect as much of my data as possible, and her knowledge and understanding of the topic at hand.

Most people working in or around these systems take almost the opposite approach: the more data, the better. It doesn't seem to matter that this approach goes directly counter to an established best practice, that of *data minimisation,* where only data that has a specific purpose is collected. This practice acknowledges that collecting and holding sensitive data creates all kinds of risks, both to the holder of that data, and to the people who the data is about. So, only collecting what's needed when it's needed, then deleting it once that purpose has been fulfilled, significantly reduces those risks. And one of the riskiest types of data to collect without a purpose (or even with a purpose) – is biometric data.

Biometric data are measurements or data drawn directly from our bodies. These days, technically speaking, there's all sorts of biometric data that can be gathered, from fingerprints to iris scans, gait measurements, measurements taken of your face, or even ear lobe measurements. (Did you know the shape of your ear is unique to you?) Biometric data often features in some way in digital identification systems, because theoretically, it's all unique to specific people.

As ever, it's not that simple. People who engage in hard labour often have worn down fingerprint ridges, meaning that their fingerprints are not read correctly by fingerprint scanners. Such scanners are usually designed for the majority, and rarely consider the experiences or realities of people at the margins, or people in the minority. There are other cultural issues to take into consideration, too. Sometimes, gathering biometric data requires a physical proximity that people from certain cultures might be uncomfortable with given the circumstances of data collection – a man needing to touch a woman's face to line it up to get the iris scan just right, for example. Again, the method of data collection is often designed for and by people for whom these kinds of cultural norms are not an issue.

Despite these issues, the collection and widespread use of biometric data is becoming more and more ubiquitous. We've become used to it in modern day life in many parts of the world – in airports you might be asked to scan your fingerprints, lots of smartphones offer the possibility of unlocking by doing a quick face scan or a fingerprint touch, and in some countries you need to give your fingerprints before you're allowed to purchase a new SIM card for your phone.

This normalisation of using your body in relation to technology – no less to actually just *access* your technology – means that the two are becoming ever more intertwined. Your body is becoming part of the way that technology recognises who you are, leaving your activities forever connected to your body.

To people designing systems who want to be sure that each person is represented once and only once, biometric systems offer a lot of possibilities, assuming they operate accurately. The world's biggest refugee agencies, the United Nations High Commission for Refugees (UNHCR), or by the World Food Programme (WFP) have been using biometrically-enabled registration cards for refugees for years, and now hold databases with millions of people's biometric data.

In some ways, this demonstrates the perceived *value* of such data. Why is it important for humanitarian agencies to gather biometric data from the people they're trying to support? Shouldn't the focus be on getting them the emergency assistance they need, rather than collecting their biometric data? Advocates of biometric systems within the humanitarian context would say that those two options aren't mutually exclusive at all: they'd say, 'we're collecting biometric data *in order to* give humanitarian assistance to people in need, in a more efficient way.' But there's more to it than that.

Biometric data is unique in comparison to other identity data in that it is immutable, there's no way of changing it (though as we'll see later, people have tried) and it's permanent. While we discussed fluid identities, biometric data is the opposite, not fluid nor growing. Your fingerprints taken at age 20 will likely be similar enough to your fingerprints aged 60 that you can be 'verified' or 'authenticated' with them, meaning that a machine could likely confirm that you are the same person who gave your fingerprints 40 years earlier. There is a potential exception for children whose biometric data, if collected very young, may no longer match them as adults (eg. fingerprints taken from a 2-year-old might not show up as a match when that person is 20 years old.) But with those few exceptions, almost all of the rest of us remain indelibly tied to anything that is associated with our biometric data, forever – for the rest of our lives.

This has huge consequences in terms of privacy, and in relation to the topics discussed earlier here. Imagine your most embarrassing moment, one where you wished the ground would swallow you up – or your biggest mistake, some moment in your life that you wish had never happened. Now imagine that that moment is irrevocably associated with your very self – every time that someone scans your fingerprints or asks you for ID, that's what they see.

This is a trivial example of how and why it's so important for people to be able to control what gets associated with their body, but there are many other much more serious examples of when a certain piece of information gets permanently associated with biometric data in a way that can shape your entire life.

At the border between Niger and Burkina Faso, people's biometric data is gathered as part of the Migration Information and Data Analysis System, or MIDAS as it is known, and shared in a centralised database held in Niamey, the capital of Niger.[66] Though only the Nigerien government have access to that database, it's linked to INTERPOL's alert lists, and linked to the biometric registration database of the US Department of State. It's up to the Nigerien government who they choose to share data from that database with. As journalist Giacomo Zandonini notes in the New Humanitarian, 'for prospective migrants at the Makalondi border post, their data is likely to travel a lot more freely than they can.'[67]

Under data sharing agreements between the European Union's border police, Frontex, and various border forces within the continent of Africa, databases like these are shared. This means that when a person arrives in the European Union, Frontex might be able to access those databases, and provide extra data to back up (or not) the story of how a person got to the EU. Given the blurriness around who gets access to which databases, and who shares data with whom, it's highly unlikely that the people whose biometric data is being collected are being properly informed about where their data might go.

This data sharing could change their lives, and not for the better. Under the Dublin Regulation, a European Union regulation adopted in 2003, people are required to seek asylum in the first European country that they enter. That means that once someone arrives in a country, and their fingerprints are taken in that country, they cannot (with some exceptions) apply for asylum elsewhere, even if their family are in another country.[68]

Eric Reidy, another journalist writing in The New Humanitarian[69], notes that the consequences of the Dublin Regulation are that it serves to 'push people towards becoming undocumented and increases the psychological stress of their long and arduous journeys. It also extends the length of time it takes for refugees to receive protection, and only then in a country where they didn't want to stay.' Many people also don't know, upon landing in a new country, that this is the case – that once they've given their fingerprint, they're required to stay in that country to get asylum, which points (again) to the problems of a lack of informed consent being provided.

There's no starker example of how a single fingerprint can change the course of someone's life. The fact that this fingerprint is forever tied to a person's body, and included as part of an unfair and unjust policy which ignores vital humane considerations such as: where does the person's family live, where do they *want* to build a family within Europe? Biometrics, in this case, are part and parcel of a fundamentally unfair and unjust approach to migration which puts more emphasis on what should be an inconsequential detail (*where did you land first?*) instead of on what *should* matter: *where do you want to go?*

Amidst such a discriminatory system, bolstered by the use of biometric technology, the only way for people to exercise their agency and self-determination is to thwart that system. There are numerous accounts throughout the years of asylum seekers and refugees burning off their own fingerprints in order to facilitate their movement.[70]

Even looking at it from the perspective of countries in the European Union, where those countries closest to borders like Italy or Greece are most likely to be the ones where people land first, and thereby hold the 'burden' of managing an influx of new people – this isn't what they want. Taking care of people who are seeking asylum and arriving in the EU should be shared across European Union countries, funded by resources and money that have most likely been extracted from the very same countries people are fleeing from. Despite many of these databases and new biometrics schemes being funded by the European Union, or other countries with data protection legislation in place *for their own citizens,* there's a distinct lack of data protection put in place in other countries.

Biometric data is also particular in comparison to other types of data in other ways. I participated in research in 2019 that looked into the lived experience of refugees who had been part of biometrically-enabled digital systems. We found that many people felt their rights were violated with the very process of collection of biometric data. For example, a woman wearing a hijab who is asked to remove the hijab in conditions she's not comfortable with for her irises to be scanned. In-country researchers found that these concerns aren't taken into account, and many people they spoke to, even those who didn't have specific concerns like having to remove their hijab, felt like their dignity was being violated.

This is worth unpacking: the collection of biometric data happens as part of registration. It might be one of the first encounters with a humanitarian agency, upon whom they are dependent for vital support. That means that (usually) before assistance has been provided, a registration or counting exercise happens; data is collected from that person by the humanitarian agency; a registration card is given to the person; and that card is then used in the future to prove that they are 'entitled' to receive assistance, like food rations.

What are the consequences of this? That biometric data collection (and the associated lack of respect for people's dignity) sets the scene for how a person feels in claiming their rights. Knowing that the institution or agency sees fit to give you food and water (and indeed, that you're dependent upon this institution for those purposes) but recognising that they don't fundamentally respect your dignity – how might that make people feel? Research has shown that when it comes to technology, and more specifically to digital identification systems, *how* people feel about the data collection is just as important as what data is collected.[71] When it comes to biometric data being taken, with a lack of thought and respect for someone's dignity and preferences, the whole process contributes to a removal of agency from people and makes the stark power dynamic very clear from the get go.

It's somewhat ironic that 'dignity' features heavily in humanitarian guidelines and principles. Researchers Irina Mosel and Kerrie Holloway note that 'dignity is a frequently invoked concept in humanitarian action', and their findings include that, for people on the receiving end of aid and assistance, dignity has more to do with '*how* aid is given, rather than *what* was given'; that transparency and fairness are key components of a dignified response; and that dignity is deeply contextually specific.[72] Suffice to say, the lack of transparency, and the *how* in these contexts do not point towards dignity being respected in the way that biometric data is gathered and used from people receiving assistance.

We've probably all been in the situation where we're asked to do something that we're uncomfortable with, and somewhere in that calculation – *Do I say yes, or do I say no* – is an implicit assessment of the power play at hand. If a woman who doesn't feel comfortable giving her fingerprints says 'no' in that biometric registration process, she goes against the wishes of the people who are providing her, and possibly also her family, with food, shelter and water. With stakes as high as this, it's barely a choice.

This is also why the idea of these agencies asking for consent from people they collect biometric data is fundamentally impossible. Nobody can be expected to 'give informed consent' when they're seeking urgent humanitarian assistance. If you're hungry, without food, thirsty for water, without a home or don't know where you and your family are going to sleep that night, the very idea that you could make an informed decision about something as abstract as your biometric, personal data, is more a cruel joke than it is the legal basis for processing data.

However, there are some cases in which collecting biometric data can be really helpful. One such example could be in the distribution of vaccines. Consider this scenario: say there's a vaccine uniquely powerful and helpful if given just once, but which reduces in effectiveness when received multiple times by the same person. In a context where mis/disinformation spreads, where people are inaccurately told multiple vaccines create extra protection against its target, it would be very helpful to be able to tell if someone had in fact received a vaccine already. Given the context, people might be incentivised to lie if asked, as they could believe that receiving multiple vaccines gives them more protection. So using biometric data is a more accurate way of checking, are the fingerprints of the person in front of me already in this database of vaccinated people, or not? It's very clearly in the best interests of the people to whom the biometric data belongs. They're the ones that need the best possible coverage from a vaccine, but who could put that immunity at risk by obtaining multiple vaccines. In theory, once the vaccine campaign has been completed and everyone has received one shot, once rates of the illness in question have gone down, one could delete the entire database of biometric data. There aren't multiple uses planned for the same data, or data sharing, merely verifying if someone already appears in this database, or not. It doesn't even really matter what their name is or any other information. It's a specific, single use of biometric data, which abides by the best practice we talked about earlier: *data minimisation.*

Unfortunately, data minimisation doesn't prevent data inaccuracy and technological error. Researcher Katja Lindskov Jacobsen analysed the roll out of biometric systems on the border of Afghanistan and Pakistan by UNHCR and showed that it was plagued with errors. Prior to being used on that dusty border, the biometric technology in question had only been tested in airports, crucially different physical conditions. Humanitarian workers trained in using the biometric finger scanners weren't warned of any potential errors, instead encouraged to trust the machine to be right. In this case, biometric data was used to signal whether someone was crossing the border for the first time, in which case they were entitled to a food and aid package. If they were re-crossing the border, they were not entitled to such a package. The question of whether not it is humane to deny people food when they or their families are starving is not considered. You can probably see where this is going.

When fingerprints matched, it was assumed that they were re-crossing and did not deserve the food. It didn't matter what people said, whether they claimed there were errors or even if they tried to show analogue proof that they hadn't crossed the border yet. Staff untrained in critically assessing the outputs of a machine, as Jacobsen reports it, stayed firm that the machine didn't lie.

While carrying out research for Oxfam in 2018, I spoke to a range of humanitarian agencies to better understand how biometrics were featuring in their work.[73] What shocked me wasn't that integrating biometrics were becoming an established part of their work, but the lack of critical thinking that was applied to that integration, and how it was the underlying policy that was the problem, not necessarily the technology or its operators.

One of the biggest arguments for the introduction of biometric data into the humanitarian sphere is that it reduces fraud. But what does 'fraud' look like here, given the specificities of how humanitarian assistance works? People I've spoken to have told me how in some extremely under-resourced and under-funded emergencies, families might only be assigned food packages that meet 50% of the daily calories they need to survive. In these cases, it's considered to be fraud if a family attempts to get two food packages for themselves, no matter the state of the recipients' health or age. This blew me away. This surely can't be classified as fraud, this is *survival.*

Humanitarian workers I spoke to agreed with me for the most part. They acknowledged that it wasn't a fair system but emphasised that it was their job to enforce said policies. They flagged a chronic lack of funding as the underlying issue, one which is not going away anytime soon[74]. They also noted that more conventional 'fraud', such as food deliveries going missing or turning up on the black market for higher prices, wasn't being addressed through the integration of biometric data.

Those involved in these more 'upstream' parts of humanitarian aid, like truck drivers or logistics personnel, are not included in this data gathering effort. This means the problem addressed through massive data gathering of refugees' and asylum seekers' data isn't the most impactful to consider, but it is the one that ensures that the humanitarian agencies hold the largest amount of biometric data.

As one former UNHCR worker told me, 'control the biometric data, control the crisis.' Biometric data holds such value for governments that if a humanitarian agency controls that valuable database, that agency inherently has more negotiating power with that government, made even more important because humanitarian agencies are only allowed to operate in countries with the agreement of the host government. This opens up some tricky debates, not least because many crises take place in countries without meaningful data protection regulation (at the time of writing, data minimisation is enshrined within European law, but not elsewhere) – and because international organisations involved in these humanitarian efforts are not usually subject to national laws in countries in which they operate. Returning to the best practice of data minimisation, if a government wants to gain access to a database after a specific purpose has been completed, the humanitarian agency in question are no longer following data minimisation principles. So with whom does the responsibility lie to ensure that biometric data collected for the purposes of refugee registration are not used for other purposes, or shared further? Refugee agencies would say they're unable to hold that responsibility because they're at the mercy of the host governments who, sadly, have been proven again and again to not have the best interests of refugees at heart.

This is true to the extreme in the case of the Rohingya. As we discussed in Chapter 3, their biometric data was collected by UNHCR in partnership with the Bangladeshi government. UNHCR assured everyone that that their biometric data would *never* be handed over to the Myanmar government, who were responsible for the violence against them in the first place. The Rohingya themselves were assured that the data they were handing over would be kept secure and never shared with the Myanmar government, too, as per the terms of their strike in November 2017.

But in 2021, Human Rights Watch revealed that in fact, the data that UNHCR had gathered had indeed been handed over to the Myanmar government[75], the very same government who led a genocide against them.

Despite this betrayal of trust – not to mention of best practices, values, and policies – the lines of accountability when it comes to handling of biometric data are so blurry that it seems like there are actually very few consequences of this incident. Observing that incident as a researcher, I thought that it would be a watershed moment, that this would be the scandal that changed the way that biometric identity data was gathered, managed, used, and shared.

I was wrong. Nothing seems to have changed following a mild media frenzy and, as far as I know, the same databases exist and are actively used. This is not okay. Zero consequences demonstrates a serious problem in the humanitarian system. As researcher Anja Kovacs highlights, 'the line between our physical and virtual bodies is increasingly becoming irrelevant…in the digital age, bodies and data are closely intertwined.'[76] The way we treat biometric data has real consequences on how people feel, the trust they've likely lost in the humanitarian system, and how humanitarian agencies are able to provide assistance in the future.

It points, once again, to the need for practising data minimisation and actually sticking to it. We need clear lines of accountability and responsibility established and so fundamental shifts need to happen within the humanitarian system at all levels. Instead of prioritising registration and processing speeds that favours data required for donors, or for governmental and agency statistics, a process that allows for the complexities of lived experience to be taken into account needs to be used. That means allowing elaborations such as 'no, I don't want to give my biometric data right now, but I *do* want to be able to access these services'. Or if a fingerprint scanner misidentifies a subject who has worked so hard for decades that their hands aren't legible to a particular system, that alternative proof is allowed and cross-referenced to demonstrate who they are, no matter what a machine declares. More focus should be put on what *people* say and need, instead of organising categories so humans are easier to process by machines. This might mean slowing down and operating less efficiently for a period of time but if it enables greater dignity and personhood, it's more than worth it.

Chapter 6
Our data-decided futures

The paths available to us in professional and personal life are so often entwined in the data surrounding us. We often don't even know that paths are made and closed to us because of it, or that such data is being gathered that can have huge effects on our potential. Such systems are often called 'black boxes', where the input is processed in a way we can't understand or see, to produce an output. It's an appropriate metaphor, as with some systems, we literally can't see what happens within the system to turn the input into whatever results the machine spits out. That's particularly problematic when the 'output' is disagreeable. Not knowing exactly what to argue with in the process, or who's at fault for a flawed or undesirable result, is a huge problem.

The city of Rotterdam, the second biggest city in the Netherlands, adopted a specially-designed machine learning model to decide who amongst their population deserve welfare payments. Newsroom Lighthouse Reports investigated the machine learning model for over two years, and their findings are truly shocking. They discovered that Rotterdam 'systematically generates suspicion of vulnerable groups', noting that 'some of the variables that increase a person's risk score are totally beyond their control: their age and gender for example.' And some variables seem to 'ethnically profile people based on the languages they speak or their ability to speak Dutch, which is widely considered a proxy for ethnicity.'[77] The system ranked every recipient of benefits on a list with those ranked in the top 10% referred for further investigation, some of whom had their benefits suspended during the investigation.

What calculated these apparent league tables? Usually, it's a combination of people and machines, or 'algorithms' that are responsible for these results, even though understanding the existence of an algorithm that doesn't make it any easier to protest yourself against it, as we see in social media advertising and categorisation. At its simplest, an algorithm is 'a series of steps, (or set of rules) for solving or performing a task.'[78] It gets more complicated when machine learning is involved. Machine learning is 'a branch of artificial intelligence in which a computer generates rules and predictions based on raw data that has been fed into it.'[79] It's important to remember, though, that people create machine learning systems, and people create the data from which the machine learning algorithms actually *learn* – meaning that if the data contains biases (which it almost always does), then the machine learning algorithm is also going to learn from, and, crucially, enforce those biases.

This kind of automated system as adopted by Rotterdam, where data makes life-changing (and, arguably, life-ending) decisions based on human and automatic biases, is becoming more common across the world. UN Special Rapporteur on poverty and human rights, Philip Alston, calls this the digital welfare state, saying 'systems of social protection and assistance are increasingly driven by digital data and technologies that are used for diverse purposes, including to automate, predict, identify, surveil, detect, target and punish.'[80]

Political scientist Virginia Eubanks goes into detail about how poor people are profiled with the use of technology, in her book *Automating Inequality*. She examines how multiple cities and states in the United States use data to make decisions that affect lives, without people knowing nor having any input into those decisions. Crucially, her book highlights how incredibly arbitrary these decisions are – applications for food stamps in Indiana denied because a computer system interprets mistakes as 'failure to cooperate', for example. Systems such as these use data that you might not expect to be considered in welfare decisions. In France, 'data from social networks can be scraped to feed machine learning algorithms that are employed to detect tax fraud.'[81] And we only know this because of the work of journalists and civil society, who work to bring *some* transparency to these opaque boxes.

One of machine learning's most insidious outcomes is arguably that data about *someone else* can be used to decide elements of *your* future. That's simply the way that machine learning works: computers find patterns in data given to them, whether by other computer-generated data or human, whether accurate or not, whether biased or not. Once the system recognises those patterns, they 'predict what is likely to happen next, assuming the pattern stays the same.'[82] That last phrase is crucial: 'assuming the pattern stays the same'. When it comes to, say, predicting the likelihood of rain given data on cloud cover, sun, and air pressure, we'd expect patterns to stay more or less the same. Unless we're talking about weather warnings for storms, natural disasters, or the climate crisis, it's unlikely that lives will be fundamentally changed as a result of this data.

When it comes to data about *people* and their behaviour, the very design of machine learning algorithms means that a pattern drawn from how *other people* behave is used to predict how *you* will behave. And because of the ways that these algorithms have been so deeply integrated into our societies, those predictions might then change what paths are available to you in life. Not because of something you intentionally did, potentially not even because of something you did in relation to the outcome in question – but just because you acted in a way that matched a pattern of behaviour of people who came before you.

In the UK in 2020, due to disruption to summer exams from the COVID-19 pandemic, the British government attempted to use an algorithm to determine A-Level grades for students who had been unable to take the exams. When the results of the algorithm were released on a student-by-student basis, almost 40% of the grades previously assigned by teachers were downgraded, meaning thousands of students would miss out on a university place that was assumed given following their own prior academic performance.[83] The algorithm used took three inputs into account, one being that each student was ranked within their own school for a particular subject. As Daan Kolkman explains, 'if a student was halfway down the ranking list, then her grade would be roughly equal to what the person in the same ranking obtained in previous years.'[84] This means that a person's grade was influenced not by their own teacher who knew that student's likelihood of getting a grade, nor by their own abilities. It meant that if 2019 students performed poorly at a specific subject, the grade assigned to a student in 2020 would be negatively impacted, even if their own teachers knew they deserved better. Even Gavin Williamson, then-Education Secretary noted 'there is sometimes a danger where you have an exceptionally high-performing child in a low-performing school to be in a situation where they don't get the grades that they want to.'[85]

This algorithm meant that not only were students' self-determination denied, but structural inequalities were also strengthened. If a student came from a low performing school in an impoverished area, they were more likely to be assigned a lower grade regardless of their actual performance at school. The system punished those who didn't happen to match the pattern of the majority. Luckily, in this case, there was significant public outcry with protests by teenagers, parents and teachers across the UK. In this case, the system used was extremely visible, and the numbers of those affected was large enough that a U-turn was obtained. In the end, teacher's predictions were used to assign students their grades and the algorithm was left unused.

As we have seen, however, most of the time, these systems aren't as visible. Germany-based non-profit organisation AlgorithmWatch has documented case upon case of how automated decision-making models are used just in Europe[86] – such as an algorithm in

Spain making risk assessments on survivors of domestic violence and whether it was likely to happen again.[87] Those more vulnerable within society are so often subjects of secretive systems and work has to be done to uncover them. In 2019, I was a co-lead researcher on a paper for UNICEF in which we examined the human rights impact of predictive analytics on children's rights.[88] We narrowed 'predictive analytics' down to meaning predictive machine learning – that is, systems that use data from the past to make inferences and predictions about the future.

We found that, for children, it's not their *own* activities affecting the opportunities available to them. In many countries, machines made decisions based on factors that were entirely outside of a child's control, like what neighbourhood they lived in, their family situation (for example, if they had relatives living with them that were known to the police), or their physical proximity to other families that had had involvement with child protective services. Based on those factors, certain inferences and decisions were made about the children themselves, what kind of support they might need, how they might behave and therefore what kind of interference and frequency of check-ups they would get from the police.

And, of course, the usual, deeply unjust, racist, anti-Black discriminatory biases play into this. In predominantly Black neighbourhoods in the United States, children are more likely to be flagged to welfare services than in white neighbourhoods, even when there is no difference in data aside from location. Similarly in the UK, children as young as 12 years old have already been entered into the Gangs Matrix, thus applying a label and identity that might have very little to do with their own actions, and much more to do with the circumstances into which they were born, their family members, their skin colour and who they associate with. Documents released by the UK Met Police also note that the data of relatives of people included in the Matrix are also collected, which may also include children or minors.[89]

This is a violation of a child's right to self-determination, which is in fact a fundamental right enshrined in the Declaration of the Rights of the Child,[90] signed and ratified by every single United Nations member state except for the United States. By making assumptions on who a child will be based on everything but their own activities, these systems unfairly remove or restrict certain opportunities from children.

But more importantly, by using automated systems like this, data is created about who a child is – or at least, who someone thinks that child is. This data might become a permanent record with welfare or protective services for a child, meaning that later on in their life if they get in trouble with the police, the two pieces of data might be put together to create a picture of the child that doesn't do justice to who they really are. It's in this way that data is used by so many systems to influence the paths available to us; and to shape our futures, whether we like it or not.

The future might look bleak thanks to such systems, but governments collecting information about their populations is nothing new. From the first tax system of Ancient Egypt around 3000 – 2800BC, to the first official UK census in 1801, humans have been tracked and conditioned. But digitisation revolutionised such processes in ways that are almost too far-reaching to comprehend. For one, digitising systems brings a scale that goes far beyond what was possible with paper identification systems. Data can be gathered, processed, stored and shared at a speed and scale that would have been unimaginable in the mid-1900s. Indeed, this is part of the reason why asking for 'consent' to store and share data from people who aren't familiar with how fast and easily data can be shared and spread, feels deeply disingenuous. Easy access to digital databases compared to analogue means it can often be almost impossible to know who has access to what, and who has copied what. Information that you give to one digital entity might be spread to another without you ever knowing. For example, current draft legislation would allow the EU to take a more 'information-driven' response to border management and security by accessing data that is gathered on the border of West African countries. The fingerprint and facial image data gathered would allow Frontex, the EU border guard, to monitor who is trying to migrate from West African countries to Europe, and to identify to which countries they can be returned. As we've seen, just having your fingerprint appear in a database can be truly life-changing, and in worst cases, life-destroying.

This introduces the challenge of permanence. Once your data is in a database of gargantuan size, it's near impossible to have it removed (but it's worth noting that the EU's GDPR Act of 2018 includes the ability to request data removal, known as the 'right to be forgotten',[91] though this only applies to European citizens). Not being able to remove or control how your data is used or stored, or where it's stored and how, severely restricts your ability to self-identify, because your entry in a database might tell a different story. Beyond individual entries in that database though, once a database containing lots of people's personal data even exists, the idea of ever considering *deleting* such a database seems to go against a lot of unwritten rules in today's political climate, in so many different contexts.

There are also significant security and privacy challenges thanks to the digitisation of all this data. Nothing is 100% secure, there is only mitigation, and we've seen over and over again (and we will continue to do so) that even the very best efforts to make sure that data is as secure as can be can be foiled at times. Sometimes it's because of malicious actors trying to access the data through any means necessary, or incompetent security policies (such as in 2018, when the personal data of 1.1 billion citizens in the Aadhaar database were being sold online for just 500 rupees),and sometimes it's because of broken promises or policy changes.

Above all, though, collecting and digitising data about people's identities in the ways that we've discussed brings a level of legitimacy to what should be entirely illegitimate processes, processes that should not exist at any level. As I'm writing this, there's a push in the UK to introduce mandatory photo ID at local elections across England, supposedly to stop identity fraud or polling station impersonation, but as *The Guardian* reported in 2018 following a trial of compulsory voter ID in 2018, 'none of the five English boroughs taking part in a controversial trial of compulsory voter ID in local elections on Thursday has experienced a single instance of polling station impersonation in the past decade.'[92] The hurried push to introduce compulsory ID seems more likely to be a way of disenfranchising already marginalised people from voting, which goes against the very premise of a democracy.

Layering technology on top of unfair policies makes them harder to reverse, and society's inbuilt technological solutionism and unbridled (and undeserved) faith in technology makes it harder for many to question the results coming from such systems. We've integrated systems that process and use data about us into crucial parts of our lives and we've done so in a way that doesn't allow for context or understanding exceptions for people who might not fit those patterns. Governments have partnered with the private sector to build those systems, giving away people's private data in the process, locking themselves into expensive systems, many of which don't even work very well.[93] As a society, or even as a group of societies, we've made some serious mistakes in how we assume that our complex identities and behaviours can be reduced to digital data points, and processed as though we're all the same, ignoring our fluid and unique histories and futures.

Conclusion

Defining ourselves for ourselves

> 'My fullest concentration of energy is available to me only when I integrate all the parts of who I am, openly, allowing power from particular sources of my living to flow back and forth freely through all my different selves, without the restrictions of externally imposed definition.'
>
> – Audre Lorde[94]

By now, we've seen many examples of how technology gathers sensitive personal data about who we are to capture our identities; and how that data then gets used to limit or change the choices available to us. Those changes affect our journeys in life, what we can and can't do, and what opportunities we have open to us. There's a variety of reasons for this, from how we think about the data itself, to the oversized importance we've given such data and digital systems in our societies today. That might seem daunting but the good thing is that means there's multiple opportunities for intervention, resistance and reclaiming our agency.

We currently have false expectations of what data can do, to the degree that we often don't even test our assumptions. Many systems that gather lots of data about our behaviour are attempting to optimise for efficiency, trying to make operations happen faster or reduce 'human cost', i.e. minimising the need for human intervention in a system. While it's the intention, it's not always the reality.

In the case of the humanitarian organisations who had been investing in biometric systems, 'increasing efficiency' was the number one motivation that they mentioned, but not a single one had any evidence to prove (or disprove) this, instead relying on assumptions and anecdotes. It was a shocking admission, given the amount of staff time and money invested in integrating these new systems, not to mention the potential privacy and trust costs of gathering such sensitive data.

Another potential route to switch things up is considering the role of data in relation to systems of power and accountability. Right now, we're gathering data to make people more visible to those in power, instead of gathering data to make power more visible to people. Effectively, accountability systems are broken. For example, in the case of humanitarian organisations, it should be the organisations being held accountable to the populations they seek to serve and support, not the other way round. People receiving aid and assistance hold the very least amount of power whereas the people providing said aid are typically handling resources, committed to using it in a certain, contracted way. Why are we taking fingerprints of those receiving food and aid? Why should those with the least power always be the ones to give up their data, instead of the most powerful?

Imagine we flipped the script on how these systems are run. Imagine if government officials were proactively and consistently held be accountable for what they're doing when they spend public funds – not just via public outcry every so often when certain egregious examples makes it into the headlines, but on an ongoing basis. Simply, people who are democratically elected in to positions of power should be held accountable for how they use that power, not recipients of social benefits having to prove why they need support or having their social media profiles monitored and, in certain cases, used against them. Government departments should have to explain exactly what an algorithm does and how it works *before* it is implemented, not after. They should have to document clearly what data about us they are gathering and define the purpose clearly, with opportunities to flag if that data is unsuitable for that purpose. Any digital system should have opportunities for redress built in – for people to say, without fear of repercussions or reprisal, if a mistake has been made.

There's also a level of critical awareness that is crucial to addressing the issues I've discussed here. Knowing that the categories we're put in are artificial and knowing that that there's a lot between the lines will hopefully mean that we're less keen to put so much stock into how technology categorises and treats us. To embrace the fluidity of our identities and allow ourselves the space to be messy, real, human people who change over time, we can acknowledge the limitations of digital systems when it comes to our identities, and do our best to hold onto that fluidity despite the rigidity of the systems we participate in.

But how?

First, let's reconsider the way that we understand our own relationships *to* technology and data. None of it is objective and 'data' involves a lot of subjectivity in terms of how it's gathered, organised, and sorted. It's important to remember that all the forms we fill in, all the boxes we tick, all the technical systems we interact with, were designed by *people* and that people have their own motivations and biases, both implicit and embedded within the design, and also explicit, motivating the very purpose for collecting the data.

Dr. Safiya Noble explores these biases comprehensively in her book *Algorithms of Oppression,* where she takes a comprehensive look at how negative biases against women of colour are embedded within the design of search engines, and other algorithmic and technical systems.[95] Similarly, Nigerian-American artist and educator Mimi Ọnụọha's work has delved into the way that absences shape our systems – the data that we *don't* collect, the way that certain aspects of collated subjects cannot be measured or quantified or datafied, and what that says or changes about the way that our systems work and our lives are shaped.

Their work, among many others, exists to encourage us all to question the systems and categories that we take for granted. In *Machine Readable Me*, I've tried to do the same, encouraging you to think twice about your understanding of personal data and how it might represent who you personally are, and who we collectively are. Data about who we are can never truly capture the intangibility of our identities, no matter how long that drop-down menu is.

And what about if we reconsider how what we value shows up in our technical systems? Ọnụọha's work has also explored the concept of 'missing data' – 'the blank spots that exist in spaces that are otherwise data-saturated,'[96] noting 'there are advantages to non-existence.'[97] If we, as a society, were willing to recognise those advantages, we could use them. In the case of identification and identity, a lack of data collection – an intentional data absence – could actually signify an intentional respect and acknowledgement that not everything can be digitised. In this day and age, innovation has become synonymous with digitisation. But what if we had a more nuanced approach to innovation? What if the most innovative approach would be to recognise that not everything should be digitised, and to understand the limitations of data gathering and of technology?

In 2019, I co-authored an article with researcher Gabriela Ivens about the ethical challenges arising from open source investigations for human rights, which is where people use data and information they find online as part of human rights research.[98] We came to the conclusion that just because people *can* find myriads of data online, doesn't mean they *should* always use those methods or that data. I believe the same about governmental systems and data about who we are. Just because a state-run system like a welfare or benefits system could theoretically integrate a machine learning algorithm doesn't mean it should. Just because data about our behaviour *could* be gathered, or even used, doesn't mean it should be. For example, social welfare systems *could* incorporate social media data automatically into how they assess someone's eligibility to receive benefits or they could rely upon what that person presents as the evidence in their favour. And in the long run, leaving that human intervention in, valuing context and lived experience of the people in question, may well result in more accurate results, and in other positive outcomes that are perhaps less easy to measure.

Like trust, for example. There are countless examples of how data about us has been misused, how systems have resulted in unjust outcomes, how people's lives have been negatively impacted by data and digital systems built within governments around the world. Each time that there are negative outcomes, I would assume that people's level of trust drops in their government and trust is essential within functioning democracies. We can't only look at quantifiable metrics that are easy to measure, such as the number of applications processed by an automated decision-making system, to decide whether a system was a success or not. As ever, we have to take people's experiences into account and think about outcomes that are far less easy to measure but that have significant impacts on how our societies run in the first place.

In 2017, I was part of a team that worked with Oxfam to explore what the most appropriate way would be for them to engage with biometric technologies in their humanitarian work. That is – how might gathering fingerprints, or iris scans, help (or hinder) their mission and their goals? As part of this research, I spoke to people who had first-hand experience of engaging with these systems as humanitarian workers and those within the systems themselves. There were no overlaps in these two groups. In fact, some people responsible for implementing these systems told me that they would be very reluctant to give away biometric data in the same way that they were requesting from refugees and asylum seekers, because of the longer-term privacy implications.

I also spoke to technical experts and, importantly, people who knew Oxfam's commitment to human rights and their mission and values. Through close work with them it became clear that not enough was known about the longer term consequences of engaging with the technology – so Oxfam decided to implement a moratorium on using biometric technology in their work. They knew that if they collected biometric data then either that data was breached or leaked, or there were errors in the system, their levels of trust with communities they seek to cooperate with would be damaged in the long run. Without trust, they can't do their work but they can do their work without biometric data, as they have been for the past 80 years. They prioritised trust over data and it remains one of the most innovative ways of engaging with a new technology that I have yet to see. Instead of following the crowd (and most other humanitarian organisations) by jumping on the new tech, they demonstrated a re-commitment to their values by taking the time to thoroughly assess the technology in question, slowing down and continuing to invest in training their staff, using systems and approaches that they knew worked well.

Beyond the technicalities of understanding the limitations of data, or even of designing and integrating systems in ways that respect and prioritise our complex human identities, we also need to think about values and human rights. If we really want to make the most of digital data in ways that do not cause harm, we have to build in values and rights from the get-go – from deciding whether or not a digital system or digital data could even appropriately address a given problem in a way that aligns with a certain set of values and respects human rights. Generally speaking, the right to self-determination needs to be prioritised in any and all forms of digital ID as well as in offline/paper identification documents. This can look differently in different contexts, from letting people self-identify their gender, to providing options that feel appropriate and representative of the population in question or ensuring that the identity categories and information that is most important to the population is in fact included on their ID documentation.

Researchers have various ways of describing the violations that arise from the misuse of data about us. Digital ethnographer Tricia Wang refers to it as violations of our digital personhood, defining personhood as 'the agency to determine one's own life decisions and outcomes.'[99] Researcher Anja Kovacs thinks about it in terms of the need for boundary management, noting the importance for us all to have the 'capacity to engage in the autonomous management of our bodies, selves and lives as we see fit.'[100]

We also need to see how these kinds of exploitative regimes of creating and using data play into a bigger picture. We've seen lots of examples here of how data about people travelling is being used to control freedom of movement, including influencing where settlement or immigration is allowed. As the climate crisis grips the world, migration will only increase. This migration is caused by the behaviour of those in richer parts of the world, causing people in poorer parts of the world to lose their homes and livelihoods.

This can't go on, no matter how 'inevitable' these systems might seem. There have been many moments of huge societal transformation and 'moral revolution',[101] and there's no reason to believe that there can't be another one in our lifetimes. Put more simply, there aren't many other viable options available to us. We cannot keep building and fortifying borders and walls, gathering biometric data to police and surveil those most at risk, policing movement, destroying the environment, without literally making this entire planet uninhabitable.

Instead, we could operate from a place of solidarity, acknowledging that we're all part of a global community. Thinking about these problems from that perspective would help to acknowledge the role of racial capitalism in creating those circumstances. We need to dismantle exploitative systems, and the only way we can do that is first with awareness of their existence – and second, active rejection of the 'norms' they try to perpetuate. This can happen on tiny scales by individuals, and actually, it doesn't have to involve technology or data at all. Building community with your neighbours, for a start; thinking about our actions not just as individuals but as members of a global community; not waiting for the negative impacts of a system to reach us before we act, but acting in solidarity with those who are already persecuted by such a system.

When it comes to the issues mentioned here, by using the categories that are provided to us by those in power who want to perpetuate divides, we give some people legibility and authority while excluding others. We must stop buying into those separations. Historian and activist Aviva Chomsky's book *Undocumented: How Immigration Became Illegal*[102] thoroughly breaks down how the creation and use of the labels of 'undocumented', 'illegal immigrant', or 'illegal alien' have been fully shaped by the political context at the time to create terms that dehumanise parts of the population dependent upon the specific type of xenophobic politics at the time – for example, anti-Asian, anti-Black, or anti-Mexican, among others.

Too many of the systems rely on similar categories and separations to decide who is allowed where, who should be in which database, whose data is 'important' and whose is not. And the result is that existing inequalities are strengthened – the poor become poorer, the richer, those in power amass more power, and those without power have their data gathered and are policed for activities that would not be an issue for richer people. As Atossa Araxia Abrahamian explains in *The Cosmopolites*[103], if you have enough money, passports and citizenship are easy to purchase. She cites an astonishing figure from Bloomberg, 'investor-citizens spent $2 billion buying passports in 2014.'[104]

These double standards are rife in the space of technology and data. Rich countries often get to decide what they are and aren't comfortable with when it comes to their data and their human rights, then throw those standards out of the window when it comes to supporting such schemes in poorer countries. These inequalities are the direct result of nation states perpetuating their sovereignty, of capitalism and the elites hoarding wealth. But while technology isn't the root cause of those issues, technology is playing a crucial, and dangerous, role in strengthening those unequal structures, and in making it harder to tear down those systems. More often than not, as hopefully this book has shown, technology that gathers data on people's identities is used to further divides and to insist on man-made, artificial divides.

There's been some fascinating work being done which examines what could be done to make a digital identification systems that would uphold our human rights, civil liberties and dignity. But the core issue that needs to be addressed to make a system that works for everyone is rarely about the data or the technology but in fact about the purpose and the motivation behind the system. There's never going to be a fair or just system which enables some people to control the movement of others, nor is there going to be a rights-upholding system which facilitates certain powerful people to decide which rights of others should be violated or respected.

Instead, digital data, and digital identification systems offer us the possibility to interrogate what's actually being measured and why – to take policies and systems that we might take for granted and to actually stop and ask who this policy benefits, who gets harmed by it, and how we can make it into a more just and fair system, long before any part of the technology is actually built. We've allowed technology to be weaponised in pursuit of prejudiced and unfair political discourse – but it doesn't have to be like this.

As feminist activist, writer and poet Audre Lorde wrote, 'What we must do is commit ourselves to some future that can include each other and to work towards that future with the particular strengths of our individual identities. And in order to do this, we must allow each other our differences at the same time as we recognize our sameness.'[105] I believe that when we reduce our complex, colourful selves to data points, we lose what makes us human. When we acknowledge how diverse and vibrant we are, as people, as participants in our ecosystems, our environment and more, and we recognise what data can and can't do – we open up hopeful and joyful futures.

References

1 *A Digital ID Handbook: Strategies for Navigating Electronic Identification Systems,* Mimi Ọnụọha and Mother Cyborg, A People's Guide to Tech, 2022. p12.

2 *The Conservation of Races,* W. E. Burghardt Du Bois. The American Negro Academy Occasional Papers, No.2. Washington, D.C.: Published by the Academy, 1897.

3 *White Logic, White Methods: Racism and Methodology,* Tukufu Zuberi and Eduardo Bonilla-Silva (eds). Rowman & Littlefield, 2008. p7.

4 "Distributing biometric ID cards to #refugees in #Senegal: "At least I have an identity now. I exist" http://rfg.ee/eFkVu" @ Refugees. *X,* 22 October 2012, 4.35AM. twitter.com/Refugees/status/260388965889228802?s=20. Accessed 4 July 2023.

5 "UNHCR distributes biometric ID cards to refugees in Senegal." Mariama Mary Fall Dia, *UNHCR Stories*, 22 October 2012. unhcr.org/news/makingdifference/2012/10/508536389/unhcr-distributes-biometric-id-cards-refugees-senegal.html. Accessed 4 July 2023.

6 "Forms Matter." Lena V Groeger, *ProPublica,* 1 November 2017. propublica.org/article/forms-matter. Accessed 10 August 2023.

7 *When Victims Become Killers: Colonialism, Nativism, and the Genocide in Rwanda,* Mahmood Mamdani. Princeton University Press, 2001. p88.

8 Ibid, p101.

9 *The Ethics of Identity,* Kwame Anthony Appiah, Princeton University Press, 2005. p105.

10 "United Nations Legal Identity Agenda." *UN Legal Identity Agenda Task Force,* unstats.un.org/legal-identity-agenda/. Accessed 20 September 2023.

11 "Over 50 000 without national documents in Kavango." Petrus Muronga, *The Namibian,* 20 June 2023. namibian.com.na/over-50-000-without-national-documents-in-ka-

vango/. Accessed 8 July 2023.

12 "Nubians in Kenya Appeal for Their "Right to Existence"." *Open Society Foundations,* 16 June 2005. opensocietyfoundations.org/newsroom/nubians-kenya-appeal-their-right-existence. Accessed 8 July 2023.

13 "About Nubian Rights Forum." *Nubian Rights Forum.* nubianrightsforum.org/about-nubian-rights-forum/. Accessed 8 July 2023.

14 "Consolidated Petitions 56, 58 & 59 of 2019." *Kenya Law,* 1 April 2019. kenyalaw.org/caselaw/cases/view/172447/. Accessed 8 July 2023.

15 "Huduma Namba drive suffers setback with 84pc budget cut." Dominic Omondi, *Business Daily Africa,* 22 February 2023. businessdailyafrica.com/bd/economy/ruto-signals-end-of-huduma-namba-with-84pc-budget-cut--4132352. Accessed 8 September 2023.

16 "Kenya Government mandates DNA-linked national ID, without data protection law." Alice Munyua, *Mozilla Open Policy and Advocacy,* 8 February 2019. blog.mozilla.org/netpolicy/2019/02/08/kenya-government-mandates-dna-linked-national-id-without-data-protection-law/. Accessed 8 September 2023.

17 *Seeing like a State: How Certain Schemes to Improve the Human Condition Have Failed,* James C. Scott. Yale University Press, 1999.

18 "Berlin 1884: Remembering the conference that divided Africa." Patrick Gathara, *Al Jazeera,* 15 November 2019. aljazeera.com/opinions/2019/11/15/berlin-1884-remembering-the-conference-that-divided-africa. Accessed 8 July 2023.

19 *The Invention of the Passport: Surveillance, Citizenship and the State,* John Torpey. Cambridge University Press, 1999.

20 "My First Passport." Orhan Pamuk, *The New Yorker,* 9 April 2007. newyorker.com/magazine/2007/04/16/my-first-passport. Accessed 10 July 2023.

21 *The Thirty Names of Night,* Zeyn Joukhadar. Atria Books, 2020. p109.

22 Roth, W. D., Yaylacı, S., Jaffe, K., Richardson, L. (2020) Do genetic ancestry tests increase racial essentialism? Findings

from a randomized controlled trial. PLoS One, 15(1): e0227399. 10.1371/journal.pone.0227399.
23 Panofsky, A., & Donovan, J. (2019). Genetic ancestry testing among white nationalists: From identity repair to citizen science. *Social Studies of Science, 49*(5), 653–681. jstor.org/stable/48596145.
24 *Superior: the Rise of Race Science,* Angela Saini. Beacon Press, 2019.
25 "When our bodies become data, where does that leave us?" Anja Kovacs, *DeepDives,* 28 May 2020. deepdives.in/when-our-bodies-become-data-where-does-that-leave-us-906674f6a969. Accessed 6 May 2023.
26 "Of 42 'Hunger-Related' Deaths Since 2017, 25 'Linked to Aadhaar Issues'." *The Wire,* 21 September 2018. thewire.in/rights/of-42-hunger-related-deaths-since-2017-25-linked-to-aadhaar-issues. Accessed 6 May 2023.
27 Ibid.
28 *The Ethics of Identity,* Kwame Anthony Appiah. Princeton University Press, 2005. p66.
29 *The Utopia of Rules: On Technology, Stupidity, and the Secret Joys of Bureaucracy,* David Graeber. Melville House, 2015.
30 "ICJ speech: Suu Kyi fails to use 'Rohingya' to describe minority." Anealla Safdar & Usaid Siddiqui, *Al Jazeera,* 13 December 2019. aljazeera.com/news/2019/12/13/icj-speech-suu-kyi-fails-to-use-rohingya-to-describe-minority. Accessed 8 May 2022.
31 "UNHCR REFUGEE POPULATION FACTSHEET." *UNHCR,* July 2018. data.unhcr.org/en/documents/download/64898
32 "Full press release about the #Rohingya refugee strike in the camps." @john_hq3, *X.* 26 November 2018, 4.39 AM. twitter.com/john_hq3/status/1066914392384532480. Accessed 12 June 2023.
33 *Fresh Banana Leaves: Healing Indigenous Landscapes through Indigenous Science,* Jessica Hernandez. North Atlantic Books, 2022. p233.
34 Seltzer, W. & Anderson, M. "The Dark Side of Numbers: The Role of Population Data Systems in Human Rights Abuses,"

Social Research, Summer 2001, Vol. 68, No.2, pp. 481-513.
35 Ibid.
36 Ibid.
37 "Forged ID Documents.", *Gedenkstätte Stille Helden,* gedenkstaette-stille-helden.de/en/silent-heroes/topics/helping-the-persecuted/forged-id-documents. Accessed 21 September 2023.
38 "During World War II, we did have something to hide." Hans de Zwart, *Medium.* 30 April 2015. medium.com/@hansdezwart/during-world-war-ii-we-did-have-something-to-hide-40689565c550. Accessed 21 September 2023.
39 Ibid.
40 "Disparities in the risk and outcomes of COVID-19", *Public Health England,* August 2020. assets.publishing.service.gov.uk/government/uploads/system/uploads/attachment_data/file/908434/Disparities_in_the_risk_and_outcomes_of_COVID_August_2020_update.pdf. Accessed 21 September 2023.
41 Hoffman, K. M., Trawalter, S., Axt, J. R., & Oliver, M. N. (2016). Racial bias in pain assessment and treatment recommendations, and false beliefs about biological differences between blacks and whites. *Proceedings of the National Academy of Sciences of the United States of America, 113*(16), 4296–4301. doi.org/10.1073/pnas.1516047113
42 "Updating ethnic contrasts in deaths involving the coronavirus (COVID-19), England: 8 December 2020 to 1 December 2021.", *Office for National Statistics,* January 2022. ons.gov.uk/peoplepopulationandcommunity/birthsdeathsandmarriages/deaths/articles/updatingethniccontrastsindeathsinvolvingthecoronaviruscovid19englandandwales/8december2020to1december2021. Accessed 21 September 2023.
43 Pareek, M. et al (2020). Ethnicity and COVID-19: an urgent public health research priority. *The Lancet.* doi.org/10.1016/S0140-6736(20)30922-3
44 "Over-Exposed and Under-Protected: The Devastating Impact of COVID-19 on Black and Minority Ethnic Communities in Great Britain." Zubaida Haque, Laia Becares and Nick Treloar, *The Runnymede Trust.* August 2020.

assets.website-files.com/61488f992b58e687f1108c7c/61c-31c9d268b932bd064524c_Runnymede%20Covid19%20Survey%20report%20v3.pdf.

45 Ibid.

46 Afrozensus, afrozensus.de. Accessed 21 September 2023.

47 "Diversity in German science: researchers push for missing ethnicity data." Hristio Boytchev, *Nature,* 5 April 2023. nature.com/articles/d41586-023-00955-9. Accessed 21 September 2023.

48 Michael Burleigh & Wolfgang Wippermann, *The Racial State: Germany 1933–1945,* Cambridge University Press, 1991. p59.

49 Ibid. p59.

50 "New Evidence that Biometric Data Systems Imperil Afghans.", *Human Rights Watch,* 30 March 2022. hrw.org/news/2022/03/30/new-evidence-biometric-data-systems-imperil-afghans. Accessed 21 September 2023.

51 *How To Be an Antiracist*, Ibram X Kendi. One World, 2019. p54

52 "The Prison Industrial Complex.", *Critical Resistance*. criticalresistance.org/mission-vision/not-so-common-language/. Accessed 21 September 2023.

53 "What is the Gangs Matrix?", *Amnesty International UK*. 18 May 2020. amnesty.org.uk/london-trident-gangs-matrix-metropolitan-police. Accessed 21 September 2023.

54 Ibid.

55 "The London Knife Crime Strategy 2017.", Mayor of London, March 2017. london.gov.uk/programmes-strategies/mayors-office-policing-and-crime/governance-and-decision-making/mopac-decisions-424. Accessed 21 September 2023.

56 "Reclaiming our Data.", Tawana Petty, Mariella Saba, Tamika Lewis, Seeta Peña Gangadharan, Virginia Eubanks, *Our Data Bodies*, 1 June 2018. odbproject.org/wp-content/uploads/2016/12/ODB.InterimReport.FINAL_.7.16.2018.pdf. Accessed 21 September 2023.

57 *To Save Everything, Click Here: The Folly of Technological Solu-*

tionism, Evgeny Morozov. PublicAffairs, 2014.
58 "Gangs violence matrix.", Metropolitan Police. met.police.uk/police-forces/metropolitan-police/areas/about-us/about-the-met/gangs-violence-matrix/. Accessed 21 September 2023.
59 "Fury in Manchester as black teenagers jailed as result of Telegram chat.", Helen Pidd, *The Guardian,* 1 July 2022. theguardian.com/uk-news/2022/jul/01/fury-in-manchester-as-black-teenagers-jailed-as-result-of-telegram-chat. Accessed 21 September 2023.
60 "About", Kids of Colour. kidsofcolour.com/about/. Accessed 21 September 2023.
61 "Germany Introduces New Gender Recognition Procedure." Kristen Sibbald, Human Rights Watch, 21 October 2022. hrw.org/news/2022/10/21/germany-introduces-new-gender-recognition-procedure. Accessed 21 September 2023.
62 Ibid.
63 Gender-concordant identity documents and mental health among transgender adults in the USA: a cross-sectional study. Scheim A.I., Perez-Brumer A.G., Bauer G.R. (2020) *The Lancet Public Health*, 5(4), pp. e196-e203.
64 "Legal recognition of your preferred gender." Citizens Information. citizensinformation.ie/en/birth-family-relationships/legal-recognition-of-preferred-gender/. Accessed 21 September 2023.
65 "Uruguay approves ground-breaking trans rights reform." Ella Braidwood, *Pink News*, 22 October 2018. pinknews.co.uk/2018/10/22/uruguay-trans-rights-law/. Accessed 21 September 2023.
"LGBTIQ+ rights." *UNHCR.* help.unhcr.org/uruguay/genero-y-diversidades/derechos-personas-lgbtiq/. Accessed 21 September 2023.
66 "Biometrics: The new frontier of EU migration policy in Niger." Giacomo Zandonini, *The New Humanitarian*, 6 June 2019. thenewhumanitarian.org/news-feature/2019/06/06/biometrics-new-frontier-eu-migration-policy-niger. Accessed 21 September 2023.
67 Ibid.
68 "How a fingerprint can change an asylum seeker's life." Eric

Reidy, *The New Humanitarian*, 21 November 2017. thenewhumanitarian.org/special-report/2017/11/21/how-fingerprint-can-change-asylum-seeker-s-life. Accessed 21 September 2023.

69 Ibid.

70 "Dublin regulation leaves asylum seekers with their fingers burnt." Harriet Grant, John Domokos, *The Guardian*, 7 October 2011. theguardian.com/world/2011/oct/07/dublin-regulation-european-asylum-seekers. Accessed 21 September 2023.

71 "Understanding the lived effects of digital ID systems." Laura Guzman, *The Engine Room*, 23 January 2020. theengineroom.org/understanding-the-lived-effects-of-digital-id-systems/. Accessed 21 September 2023.

72 "Dignity and humanitarian action in displacement." *HPG*, March 2019. cdn.odi.org/media/documents/Dignity_synthesis_paper.pdf. Accessed 21 September 2023.

73 "Biometrics in the Humanitarian Sector." *The Engine Room*, 2018. theengineroom.org/wp-content/uploads/2018/03/Engine-Room-Oxfam-Biometrics-Review.pdf. Accessed 21 September 2023.

74 "With needs at a record high, underfunding is chronic Guterres tells humanitarians." *UN News*, 21 June 2023. news.un.org/en/story/2023/06/1137972. Accessed 21 September 2023.

75 "UN Shared Rohingya Data Without Informed Consent." *Human Rights Watch*, 15 June 2021. hrw.org/news/2021/06/15/un-shared-rohingya-data-without-informed-consent. Accessed 21 September 2023.

76 "Towards an embodied approach to data." *Internet Democracy Project*. internetdemocracy.in/bodies-and-data. Accessed 21 September 2023.

77 "Suspicion Machines Methology." Justin-Casimir Braun, Eva Constantaras, Htet Aung, Gabriel Geiger, Dhruv Mehrotra, Daniel Howden, *Lighthouse Reports*. lighthousereports.com/suspicion-machines-methodology/. Accessed 21 September 2023.

78 "A People's Guide to AI." alliedmedia.org/wp-content/uploads/2020/09/peoples-guide-ai.pdf. Accessed 21 September 2023.

79 Ibid.

80 "World stumbling zombie-like into a digital welfare dystopia, warns UN human rights expert" UNHR, 17 October 2019. ohchr.org/en/press-releases/2019/10/world-stumbling-zombie-digital-welfare-dystopia-warns-un-human-rights-expert?LangID=E&NewsID=25156. Accessed 21 September 2023.

81 "Automating Society." October 2020. automatingsociety.algorithmwatch.org/wp-content/uploads/2020/12/Automating-Society-Report-2020.pdf. Accessed 21 September 2023.

82 Ibid, "A People's Guide to AI."

83 "A-level results: almost 40% of teacher assessments in England downgraded." Richard Adams, Sally Weale, Caelainn Barr, *The Guardian*, 13 August 2020. theguardian.com/education/2020/aug/13/almost-40-of-english-students-have-a-level-results-downgraded. Accessed 21 September 2023.

84 ""F**k the algorithm"?: What the world can learn from the UK's A-level grading fiasco." Daan Kolkman, *LSE*, 26 August 2020. blogs.lse.ac.uk/impactofsocialsciences/2020/08/26/fk-the-algorithm-what-the-world-can-learn-from-the-uks-a-level-grading-fiasco/. Accessed 21 September 2023.

85 Ibid, "A-level results: almost 40% of teacher assessments in England downgraded."

86 Ibid, "A People's Guide to AI."

87 "The first protective measures for an abused woman are decided by a computer algorithm." Jose Precedo, elDiario, 23 September 2016. www.eldiario.es/sociedad/primeras-proteccion-maltratada-algoritmo-informatico_1_3824929.html. Accessed 21 September 2023.

88 "Predictive Analytics for Children." Zara Rahman, Julia Keseru, UNICEF, November 2021. unicef-irc.org/publications/1275-predictive-analytics-for-children-an-assessment-of-ethical-considerations-risks-and-benefits.html. Accessed 21 September 2023.

89 Ibid.

90 "Convention on the Rights of the Child" ohchr.org/en/instruments-mechanisms/instruments/convention-rights-child. Accessed 21 September 2023.

91 "Everything you need to know about the "Right to be forgotten"." GDPR. gdpr.eu/right-to-be-forgotten/. Accessed 21 September 2023.
92 "England voter ID trial 'a solution in search of a problem'." Peter Walker, *The Guardian*, 29 April 2018. theguardian.com/politics/2018/apr/29/england-voter-id-trial-discrimination-fears-electoral-fraud
93 Ibid, "Automating Society."
94 *Sister Outsider*, Audre Lorde. Crossing Press, 1984.
95 *Algorithms of Oppression: How Search Engines Reinforce Racism*, Safiya Umoja Noble. NYU Press, 2018.
96 "The Library of Missing Datasets (2016)" Mimi Ọnụọha. mimionuoha.com/the-library-of-missing-datasets. Accessed 21 September 2023.
97 "On Missing Data Sets", Mimi Ọnụọha. github.com/MimiOnuoha/missing-datasets. Accessed 21 September 2023.
98 *Digital Witness*, Sam Dubberley, Alexa Koenig, Daragh Murray. OUP: 2020.
99 "You are not your data but your data is still you." *Deep Dives*, 7 August 2020. medium.com/deep-dives/you-are-not-your-data-but-your-data-is-still-you-b41d2478ece2. Accessed 21 September 2023.
100 "When our bodies become data, where does that leave us?" Anja Kovacs, *Deep Dives*, 28 May 2020. medium.com/deep-dives/when-our-bodies-become-data-where-does-that-leave-us-906674f6a969. Accessed 21 September 2023.
101 *The Honor Code: How Moral Revolutions Happen*, Kwame Anthony Appiah. W. W. Norton, 2011.
102 *Undocumented: How Immigration Became Illegal,* Aviva Chomsky, Beacon Press. 2014.
103 *The Cosmopolites: The Coming of the Global Citizen,* Atossa Araxia Abrahamian. Columbia Global Reports, 2015.
104 Ibid.
105 *The Master's Tools Will Never Dismantle the Master's House*, Audre Lorde. Penguin Classics, 2018.

Acknowledgements

This book draws upon research I was part of from over the past 11 years – to everyone involved in all those projects, the biggest thank you. All mistakes and misrepresentations are very much my own. Much of the digital ID and biometrics research mentioned here happened while I was at the Engine Room, so a big thank you to colleagues past and present, funders, partners, and collaborators. This work also draws upon research I carried out while holding fellowships at Data & Society, digitalHKS at the Harvard Kennedy School, and Stanford University's Digital Civil Society Lab, and I'm grateful for their support.

A huge thank you to Laura Jones-Rivera and Heather McDaid at 404 Ink for giving me the opportunity to write this book, and in particular to Laura for her thoughtful edits and feedback. A special thanks to Michael Brennan and the Tech + Society team at Ford Foundation for funding a book workshop which enabled me to get feedback on early drafts of this book.

I'm so grateful to the many people who have generously talked over ideas for this book, read over drafts, and offered feedback and celebration over the years: Salmana Ahmed, Chenai Chair, Eva Constantaras, Alix Dunn, Mandy Van Deven, Gauri von Gulik, Jacqui Howard, Jac Sm Kee, Laura Guzmán, Julia Keseru, Mallory Knodel, Anja Kovacs, Solana Larson, Matthew Mahmoudi, Vidushi Marda, An Xiao Mina, Ana Polanco, Mimi Ọnụọha, Michelle Thorne, Xiaowei Wang, and the many more people who I've bounced ideas around with in the past few years – too many to name, but I appreciate you!

This vast majority of this book was written while pregnant with my daughter, and finalised with her sleeping next to me. It was an absolute joy to be able to embrace family life while writing alongside it, none of which would've been possible without my family and friends supporting me. They say it takes a village, and they're not lying: the hugest thank you for being there for me, Natasha Bonnot, Corinne Cath-Speth, Hayley Fallon-Bond, Anita Gohdes, Jenny Herriott, Gabriela Ivens, Julia Kloiber, Ling Luther, Sorcha McMahon, Chitra Nagarajan, Shirley Ogolla, Mimi Ọnụọha, Fiona Power, Sumaiya Rahman, Lavinia Schwedersky, and Laura Watson.

I'm grateful beyond words to my mum for so many things, but especially for teaching me early and often never to underestimate myself. Thank you to my family for always being there for me, and to my niece Alina Rashid for inspiring me to write.

And finally, Stefan, thank you for believing in me so absolutely, and for your commitment to making me laugh no matter how ridiculous. And my biggest loves, Ovi and Amaya. Becoming your mama showed me I'm more powerful than I ever realised – thank you for teaching me to soar, I hope I can do the same for both of you. Every day with our little family is a dream come true.

About the Author

Zara Rahman is a British-Bangladeshi researcher, writer and activist based in Berlin. She has worked in civil society for over a decade, supporting context-driven uses of technology and data in pursuit of social justice and racial equity. She has held fellowships at Stanford University and the Harvard Kennedy School, and is a trustee at Saheli, a charity providing support and refuge to women of colour fleeing domestic abuse.

Also in the Inklings series

Electric Dreams – Heather Parry

Electric Dreams picks apart the forces that posit sex robots as either the solution to our problems or a real threat to human safety, and looks at what's being pushed aside for us to obsess about something that will never happen.

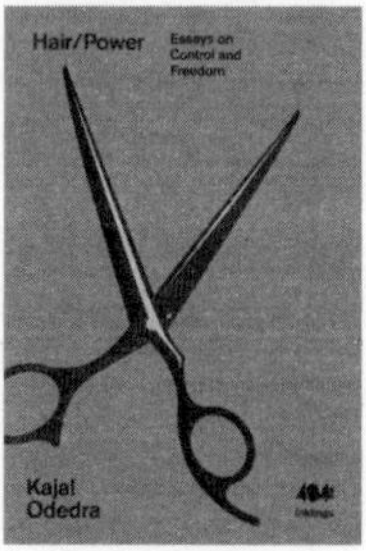

Hair/Power – Kajal Odedra

Kajal Odedra considers how hair has shaped society today, from the 'perfect' blondes in the school playground to the angry skinheads on the streets. Mohawks, wigs, afros, these are just a few of the ways in which hair has been part of history and wider activism.

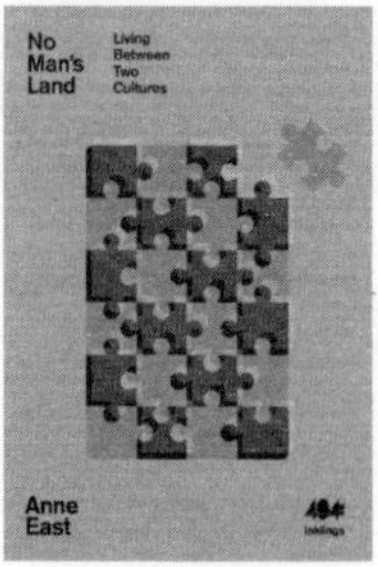

No Man's Land – Anne East

Many individuals – especially of non-white heritage – are suspended in an identity limbo. The need to divide and separate our lives, even ourselves, into neat boxes means that many British-born people with no ties to their parental culture are left adrift within our society. This is Anne East's experience.